Energy-Smart Buildings

Design, construction and monitoring of buildings for improved energy efficiency

IOP Series in Renewable and Sustainable Power

The IOP Series in Renewable and Sustainable Power aims to bring together topics relating to renewable energy, from generation, to transmission, storage, integration and use patterns, with a particular focus on systems-level and interdisciplinary discussions. It is intended to provide a state-of-the-art resource for all researchers involved in the power conversation.

Series Editor
Professor David Elliott
Open University, UK

About the Editor
David Elliott is emeritus Professor of Technology Policy at the Open University, where he developed courses and research on technological innovation, focusing on renewable energy policy. Since retirement, he has continued to write extensively on that topic, including a series of books for IOP Publishing and a weekly blog post for *Physics World* (physicsworld.com/author/david-elliott)

About the Series
Renewable and sustainable energy systems offer the potential for long-term solutions to the world's growing energy needs, operating at a broad array of scales and technology levels. The IOP Series in Renewable and Sustainable Power aims to bring together topics relating to renewable energy, from generation, to transmission, storage, integration and use patterns, with a particular focus on systems-level and interdisciplinary discussions. It is intended to provide a state-of-the-art resource for all researchers involved in the power conversation.

We welcome proposals in all areas of renewable energy including (but not limited to): wind power; wave power; tidal power; hydroelectric power; PV/solar power; geothermal power; bioenergy; heating; grid balancing and integration; energy storage; energy efficiency; carbon capture; fuel cells; power to gas; electric/green transport; energy saving and efficiency.

Authors are encouraged to take advantage of electronic publication through the use of colour, animations, video, data files and interactive elements, all of which provide opportunities to enhance the reader experience.

Do you have an idea for a book you'd like to explore?
We are currently commissioning for the series; if you are interested in writing or editing a book please contact Caroline Mitchell at caroline.mitchell@ioppublishing.org.

Energy-Smart Buildings

Design, construction and monitoring of buildings for improved energy efficiency

Edited by
Jacob J Lamb and Bruno G Pollet
NTNU, Institutt for energi-og prosessteknikk, Kolbjørn Hejes Vei 1B, Trondheim, NO-7491, Norway

IOP Publishing, Bristol, UK

ISBN 978-0-7503-3259-0 (ebook)
ISBN 978-0-7503-3257-6 (print)
ISBN 978-0-7503-3260-6 (myPrint)
ISBN 978-0-7503-3258-3 (mobi)

DOI 10.1088/978-0-7503-3259-0

Version: 20200401

IOP ebooks

British Library Cataloguing-in-Publication Data: A catalogue record for this book is available from the British Library.

Published by IOP Publishing, wholly owned by The Institute of Physics, London

IOP Publishing, Temple Circus, Temple Way, Bristol, BS1 6HG, UK

US Office: IOP Publishing, Inc., 190 North Independence Mall West, Suite 601, Philadelphia, PA 19106, USA

Contents

Preface

The European building stock is responsible for up to 36% of the CO_2 emissions. It is also stated that about 75% of the European building stock is considered energy inefficient. In order to lower the CO_2 emissions, and improve the energy efficiency of the European building stock, significant changes need to be made to the existing building stock and future buildings. More than 80% of our present energy consumption is chemical and non-renewable (coal, oil and gas). The most important renewable energy alternatives (wind and solar energy) need to be implemented to reduce CO_2 emissions. The renewable transition will require a collective effort using many different types of energy conversion and storage devices and technologies to remove the dependence on non-renewable fossil fuels entirely.

The implementation of near-zero-emission buildings (nZEBs) has become the standard for the future of the European building stock, with the development of nZEB standards increasing. The nZEBs are defined as buildings that have very high energy performance with nearly zero, or meagre amounts of energy use. The energy used in nZEBs is provided by energy from renewable sources, including sources produced on-site or nearby. One challenge is that nZEB requirements and standards are mainly targeted towards new buildings, as this is the European objective. For these standards to be fully capable, they should also provide adjusted requirements for retrofitting of the current building stock.

This volume intends to provide a brief research source for building technology and regulations in terms of energy efficiency, discussing fundamental aspects as well as cutting-edge trends for new buildings and retrofitting the current building stock. Sources of renewable and sustainable energy production and storage are additionally reviewed, with case studies of such systems on buildings in a cold climate. This volume provides industry professionals, researchers and students with the most updated review on modern building ideas, and renewable energy technologies that can be coupled with them. It aims to help readers identify technology gaps, develop new ideas and novel designs that lead to an improvement of the European building sector.

The editors and authors are grateful to the ENERSENSE* program, the ENERSENSE team and NTNU Team Hydrogen** at the Norwegian University of Science and Technology (NTNU) for supporting and helping on this book volume.

Dr Jacob J Lamb and Professor Dr Bruno G Pollet
Editors

***About ENERSENSE**

ENERSENSE

NTNU
Norwegian University of
Science and Technology

ENERSENSE is a strategic research area with a focus on the nexus of energy efficiency, energy storage and sensor technologies, including automation.

****About NTNU Team Hydrogen**

NTNU HYDROGEN

NTNU – Trondheim
Norwegian University of
Science and Technology

NTNU
Norwegian University of
Science and Technology

NTNU Team Hydrogen is a team of world experts on Hydrogen Energy. The team consists of researchers from different disciplines, departments and faculties across NTNU that works within the Hydrogen area. One of the main tasks of the team is to develop new research programs and projects both nationally and internationally with academia, research organisations and industry, and to coordinate Hydrogen activities within NTNU. We are also educating and training research leaders, innovating, providing solutions and stimulating the industry.

'Conversion into chemical energy (e.g., hydrogen, batteries and supercapacitors) allows for energy systems that supply energy in the right time, form and place.' Professor Odne S Burheim, ENERSENSE Leader, NTNU.

List of contributors

Inger Andresen
Department of Architecture and Technology
Faculty of Architecture and Design
inger.andresen@ntnu.no

Mathias Bjugan
Department of Energy and Process Engineering
Faculty of Engineering, NTNU
mathias@bjugan.no

Anne A Brajkovic
Department of Energy and Process Engineering
Faculty of Engineering, NTNU
anneabr@stud.ntnu.no

Odne S Burheim
Department of Energy and Process Engineering
ENERSENSE & NTNU Team Hydrogen
Faculty of Engineering, NTNU
odne.s.burheim@ntnu.no

Tania Bracchi
Department of Energy and Process Engineering
ENERSENSE
Faculty of Engineering, NTNU
tania.bracchi@ntnu.no

Fredrik Dessen
Department of Engineering Cybernetics
ENERSENSE
Faculty of Information Technology and Electrical Engineering, NTNU
fredrik.dessen@ntnu.no

Laurina C Felius
Department of Civil and Environmental Engineering
ENERSENSE
Faculty of Engineering, NTNU
laurina.felius@ntnu.no

Martin Gjertsen
Department of Energy and Process Engineering
Faculty of Engineering, NTNU
martin.g@online.no

Bjørnar Hamre
Department of Energy and Process Engineering
Faculty of Engineering, NTNU
bjornar.hamre@ntnu.no

Bozena D Hrynyszyn
Department of Civil and Environmental Engineering
ENERSENSE
Faculty of Engineering, NTNU
bozena.d.hrynyszyn@ntnu.no

Jacob J Lamb
Department of Electronic Systems, Department of Energy and Process Engineering
ENERSENSE
Faculty of Information Technology and Electrical Engineering, NTNU
Jacob.j.lamb@ntnu.no

Kristian M Lien
Department of Energy and Process Engineering
ENERSENSE
Faculty of Engineering, NTNU
Kristian.m.lien@ntnu.no

Bruno G Pollet
Department of Energy and Process Engineering
ENERSENSE & NTNU Team Hydrogen
Faculty of Engineering, NTNU
bruno.g.pollet@ntnu.no

Chapter 1

Introduction to energy efficiency in buildings

Laurina C Felius, Bruno G Pollet and Jacob J Lamb

Reduction in greenhouse gas emissions is essential for the development of a sustainable society. The building sector represents a large consumer of greenhouse gas emission producing energy. As a result, it is essential that the efficiency and sustainability of buildings is significantly improved to reduce the building sector's contribution to greenhouse gas emission. This chapter intends to give an introduction to the building sector and the areas within the sector that can be improved.

1.1 Introduction

The importance of energy efficiency and lowering greenhouse gas (GHG) emissions has been emphasised in recent years because of higher energy prices and increased awareness of climate change. The current climate change being witnessed is mainly caused by energy related GHG emissions from fossil fuel utilisation. Based on data from the European Union, the existing build stock currently accounts for around 40% of the total energy consumption and 36% of the CO_2 emissions in the European Union, and the building sector is still expanding [1]. Increasing the energy efficiency of the building sector is required to lower the total energy consumption and CO_2 emissions significantly. This relates to both new and existing buildings as 75% of the building stock is estimated to be inefficient in terms of energy use [1].

One strategy to design energy-efficient buildings is to follow a holistic approach, as shown in figure 1.1. There have been many studies focusing on individual steps over the past years, but a holistic overview is missing. This book aims to summarize strategies, concepts and technologies that can improve the energy efficiency of buildings in cold climates. Also, some case studies looking at the implementation of such strategies and concepts are given and discussed. The book addresses and discusses the following ideas:

- how can intelligent buildings be designed in cold climates;
- how can we evolve to a more energy efficient and sustainable building sector; and,

doi:10.1088/978-0-7503-3259-0ch1

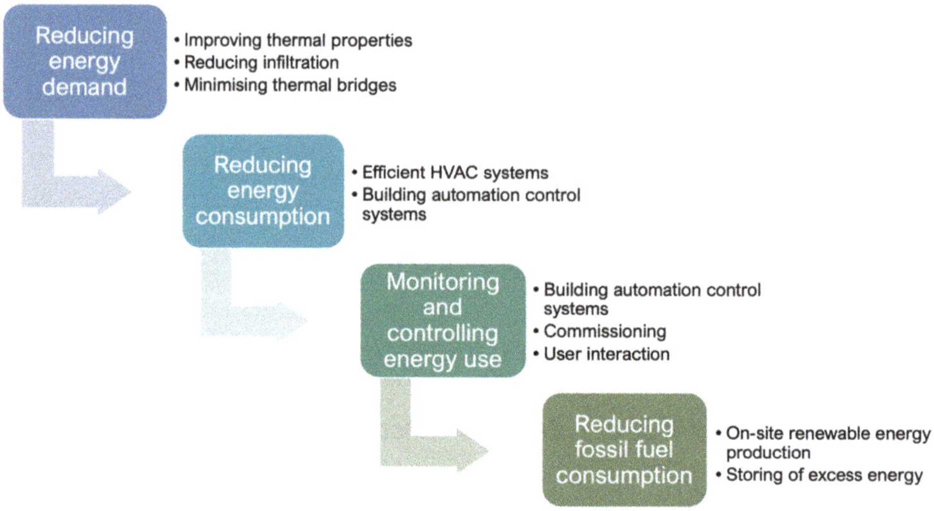

Figure 1.1. A holistic design approach towards energy efficient buildings in cold climates. Adapted from [2] with permission of Springer.

- how can the building stock in cold climate develop towards European targets and directives?

1.2 Protocols and legislation

The European Union has published several directives to address and improve the issues around the energy efficiency of the building sector and to reduce their GHG emissions. These directives were inspired by the Kyoto Protocol that was adopted in 1997 and entered into force in 2005. The Kyoto Protocol is an international agreement that sets internally binding targets for reducing emissions. It has been approved by parties of the United Nations Framework Convention on Climate Change (UNFCCC), including the European Union, and is supported by most countries. The Protocol focuses on reducing emissions of six greenhouse gases and includes binding targets for the future. During the first commitment period (period ending in 2013), the binding target for developed countries was to reduce overall GHG-emissions by at least 5% compared to 1990 levels [3]; however, each country has its own targets. The Doha Amendment was presented in 2013 and is active until 2020. This commitment focuses on stabilizing the concentration of GHG emissions in the atmosphere. The amendment states that the overall GHG emission should be reduced by at least 18% by 2020, though many countries have individual targets [4].

Since the Kyoto Protocol does not have binding targets for developing countries and many significant emitters are not part of the Kyoto Protocol, all Parties of UNFCCC adopted the Paris Agreement in 2015. This is the first legally binding global climate agreement and will be implemented from 2020. The Paris Agreement states that global GHG-emissions should be reduced by a minimum of 60% below 2010 levels by 2050 and aims to limit the average increase of the global temperature to 2 °C [5]. All countries that agreed have to commit to long-term strategies for reducing GHG-emissions. It furthermore states that the G20 nations (19 countries plus the EU), which represents 75% of global emissions, are expected to implement the Protocol from 2020. The Protocol will be reviewed every five years to set more ambitious targets for the next period based on recent developments.

The European Union has given increased importance to reaching the Paris Protocol targets and pledged to reduce GHG-emissions by at least 40% by 2030. To reach this target and to continue to be on the forefront of the global energy transition, the European Commission presented the Clean Energy for All Europeans package [6]. This package has five main targets:

- prioritize energy efficiency with an emphasis on the building sector;
- increase the use of renewables;
- improve the governance of the Energy Union;
- give consumers more rights regarding production, storing and selling their energy; and,
- improve the electricity market to be smarter and more efficient.

Out of the proposed documents in the Clean Energy Package, the Energy Efficiency Directive (EED) and Energy Performance in Buildings Directive (EPBD) are the most relevant for decreasing GHG-emissions from the building sector. They set targets for energy efficiency, energy renovation (i.e., renovation to improve the energy performance) and energy performance of buildings [7–9]. These targets include the objectives to save 20% of the primary energy consumption by 2020 through energy efficiency targets [8], to reduce overall GHG-emissions by at least 20% by 2020 [9] and by at least 40% by 2030 [7]. The Renewable Energy Directive (RED) sets ambitious targets for the production of renewable energy in the EU. This directive should ensure that renewables cover at least 32% of the primary energy consumption by 2030 [10]. The most critical points of these three directives within the scope of this book are outlined below.

Energy performance of buildings:

- The rate of renovation projects of the existing building stock needs to be increased. The central government of each Member State needs to achieve an annual renovation rate of 3% of the heated floor area of heated or cooled buildings that they own and occupy in accordance with article 5 of Directive 2012/27/EU [8];
- new and existing buildings that undergo major renovation should meet the minimum energy performance criteria, set by the Member State in accordance with article 4 of Directive 2010/31/EU [9];
- all new buildings should be nearly zero-energy buildings by the end of 2020 and all new buildings occupied and owned by public authorities should be nearly zero-energy buildings by the end of 2018 [9]; and,
- new public buildings and public buildings that undergo major renovation should fulfil an exemplary role in the context of the Directives [7–10].

Strategies and requirements:

- Member States should establish long-term strategies to mobilize investing in the energy renovation of buildings. This should include mapping the national building stock, identifying cost-effective renovation solutions, policies and measures to stimulate renovations, a perspective to guide investment decisions and an estimate of expected energy savings and other benefits [7–9];
- buildings in accordance with articles 11 and 12 of Directive 2010/31/EU should have an energy performance certificate [9];
- Member States are to specify minimum energy performance requirements and calculate cost-optimal levels for these requirements. Member States are to specify system requirements for technical building systems to optimize energy use. These systems should also be assessed when installed, replaced or upgraded [9];
- Member States are to introduce measures in their building regulations to increase the use of renewable energy resources in the building sector. Member States are to specify requirements for the minimum share of renewable energy sources in new buildings and buildings that undergo major renovations [10]; and,
- Member States should promote renewable heating and cooling systems that can significantly reduce the energy consumption [10].

Intelligent buildings:

- The consumer should be provided, if possible (see article 9 of Directive 2012/27/EU), with individual meters that monitor and inform the consumer about the actual energy consumption per energy source. Also, all consumers should have access to their current and historical energy bills and billing information free of charge [8]; and,
- member States are to ensure that new buildings are equipped with self-regulating temperature control systems in individual zones or rooms. This is also required in existing buildings when heat generators are replaced [9].

Renewable energy sources:

- All consumers can become self-consumers of renewable energy and are allowed to produce, store and sell their energy. Self-consumers located in the same building are allowed to share renewable energy that is produced on-site [10]; and,
- consumers connected to inefficient district heating or cooling systems should be allowed to disconnect when they can produce heating or cooling more efficiently from renewable resources [10].

1.3 Energy performance standards in the European Union

Requirements and standards for energy performance criteria in cold European climates are updated regularly to become stricter. The energy performance of buildings is increasingly important with the ongoing climate issues and rising energy prices. The Buildings Performance Institute Europe (BPIE) is dedicated to improving the energy performance of European buildings and ensures that European directives are implemented correctly [11]. Despite this, there are no international building codes for the energy performance of buildings. The European directives state that each country is responsible for presenting its requirements to conform with the objectives listed in the directives.

Building codes in cold climates are stricter than those in warmer climates when it comes to the energy performance and space heating demand. Scandinavian countries developed insulation requirements for buildings around 1960. The cold and airy buildings caused health problems, especially during the winter. Since then, countries have tightened their requirements towards more energy-efficient buildings. Felius *et al* (2019) gave an overview of the energy performance requirements in the cold European climates for both new buildings and renovated buildings [2]. These include Norway, Finland, Sweden, Estonia and Latvia. They concluded that what is considered as 'energy efficient' differs per country, as the energy performance criteria use different parameters to quantify the overall energy performance and significantly differ in the U-value criteria for the building envelope. These differences illustrate that building codes, whether they are on a national or international level, should account for local climate zones. It is also essential to reach the European targets to have a clear understanding of what an 'energy-efficient' building is in terms of quantified energy performance. The study also observed that not all countries have building codes adjusted for retrofitting projects. Renovating the building stock is an essential objective of the European Union but retrofitting towards the energy performance requirements for new buildings can be challenging [2].

Though there are no international building codes, there are several other incentives and standards created to improve the energy efficiency of buildings. National building codes contain the minimum requirements for energy performance, but standards can be followed to achieve more ambitious levels of energy efficiency. Several European cold climate countries have developed their low-energy or zero-energy standards. These standards are not obligatory to fulfil, though it is expected that national building codes will develop towards this level of ambition when it comes to energy performance. In particular, the development of nZEB standards is increasing, as the EPBD states that all new buildings should be nZEB by 2020. The European Commission defines this as 'a building that has a very high energy performance with the nearly zero, or very low amount of energy required covered to a very significant extent by energy from renewable sources, including energy from renewable sources produced on-site or nearby' [12]. However, each country is responsible for its definition, resulting in the fact that there is no accepted international definition [13, 14]. Another challenge for the building stock is that nZEB requirements are mostly targeted towards new buildings, as this is the European objective. As with the national building codes, nZEB standards should provide adjusted requirements for retrofitting.

An energy performance standard that is defined on an international level is the Passive House (PH) Standard. The PH-Standard focuses on reducing the heating demand of a building while increasing the thermal comfort. Both the heating and cooling demand should not exceed 15 $kWh/m^2 \cdot year$, regardless of the building type or climate zone [15]. At the same time, it may be challenging to achieve this Standard, due to location and geometry constraints and varying remaining lifetime expectancies of the components. The EnerPHit Standard is targeted towards retrofitting projects carried out at once or in steps over time. The Standard contains requirements for retrofitting projects, adjusted to the building type and climate zone, and addresses challenges related to retrofitting, especially when it is carried out in steps [2].

References

[1] European Commission Energy performance of buildings | Energy [Internet]. 2019 [cited 2019 Oct 7]. Available from: https://ec.europa.eu/energy/en/topics/energy-efficiency/energy-performance-of-buildings/overview

[2] Felius L, Dessen F and Hrynyszyn B 2019 Retrofitting towards energy efficient homes in European cold climates—a review *Energy Effic.* **13** 101–25

[3] United Nations 1998 Kyoto Protocol to the United Nations Framework convention on climate change [Internet]. [cited 2019 Oct 7]. Available from: https://unfccc.int/resource/docs/convkp/kpeng.pdf

[4] UNFCCC Doha Amendment to the Kyoto Protocol [Internet]. UNFCCC. 2012 [cited 2019 Oct 7]. Available from: https://unfccc.int/kyoto_protocol/doha_amendment/items/7362.php%5Cnhttps://treaties.un.org/doc/Publication/CN/2012/CN.718.2012-Eng.pdf

[5] Commission EU 2015 The Paris Protocol–A blueprint for tackling global climate change beyond 2020 *COM.* **81**

[6] European Commission Clean energy for all Europeans [Internet]. 2019. Available from: file:///Users/jacoblamb/Desktop/MJ0319092ENN.en.pdf

[7] European Parliament Directive (EU) 2018/844 of the European Parliament and of the Council of 30 May 2018 amending Directive 2010/31/EU on the energy performance of buildings and Directive 2012/27/EU on energy efficiency [Internet]. 2018 [cited 2019 Oct 7]. Available from: https://eur-lex.europa.eu/legal-content/EN/TXT/PDF/?uri=CELEX:32018L0844&from=EN

[8] Parliament E 2012 Directive 2012/27/EU of the European Parliament and of the Council of 25 October 2012 on energy efficiency, amending Directives 2009/125/EC and 2010/30/EU and repealing Directives 2004/8/EC and 2006/32/EC. Europenan Environment Agency

[9] Recast E 2010 Directive 2010/31/EU of the European Parliament and of the Council of 19 May 2010 on the energy performance of buildings (recast). European Environment Agency

[10] European Parliament Directive (eu) 2018/2001 of the European Parliament and of the council of 11 December 2018 on the promotion of the use of energy from renewable sources (recast) [Internet]. 2018 [cited 2019 Oct 7]. Available from: https://eur-lex.europa.eu/legal-content/EN/TXT/PDF/?uri=CELEX:32018L2001&from=EN

[11] Buildings Performance Institute Europe 2019 Improving the energy performance of buildings across Europe

[12] D'Agostino D, Zangheri P, Cuniberti B, Paci D and Bertoldi P 2016 Synthesis report on the national plans for nearly zero energy buildings (NZEBs). Publications Office of the European Union

[13] Panagiotidou M and Fuller R J 2013 Progress in ZEBs—A review of definitions, policies and construction activity *Energy Policy* **62** 196–206
[14] Sartori I, Napolitano A and Voss K 2012 Net zero energy buildings: A consistent definition framework *Energy Build.* **48** 220–32
[15] PHI PHI 2016 Criteria for the Passive House, EnerPHit and PHI Low Energy Building Standard. Passiv House Inst.

IOP Publishing

Energy-Smart Buildings
Design, construction and monitoring of buildings for improved energy efficiency
Jacob J Lamb and Bruno G Pollet

Chapter 2

Basic principles of energy use in buildings

Laurina C Felius and Jacob J Lamb

Before elaborating on technologies, solutions and strategies to improve the energy efficiency of a building, it is essential to know which elements impact the energy consumption of a building. This chapter gives an overview of the basic principles of heat transfer and the energy balance of a building.

2.1 Basic principles of heat transfer

Temperature differences cause heat transfer from the highest temperature to the lowest. All heat transfers consist of one or a combination of the three main heat transfer mechanisms: conduction, convection and radiation (figure 2.1). How the heat is transferred and how the total heat flow is calculated is explained in this section. Also, heat can be transferred through evapotranspiration and when a phase change of a substance happens (e.g. ice melting to liquid water).

2.1.1 Heat conduction

Conduction is the heat transfer through the material, which can be a solid, liquid or gas, by the movement of molecules in the material. When molecules with a different temperature collide, heat is transferred in the form of kinetic energy. Heat conduction occurs in every material, though the effect is affected by the thickness of the layer, the thermal conductivity and the temperature difference. Fourier's law equation (2.1) says that the heat flow, dQ, through a homogenous material with the area, dA, during the period, dt, is proportional with the temperature gradient, $(\delta T/\delta x)$, in the opposite direction [1, 2]. In other words, the magnitude of the heat flow is proportional to the temperature difference and in accordance with the second law of thermodynamics, it is tough to flow in the opposite direction (i.e. heat flows from the highest to the lowest point). A flat and homogenous layer with fixed temperatures on the interior and exterior has a density of heat flow rate, q_l, according to equation (2.2). In this equation, λ is the thermal conductivity of the material in W m^{-1} K^{-1} and d is the thickness of the layer, perpendicular to the heat flow, in m.

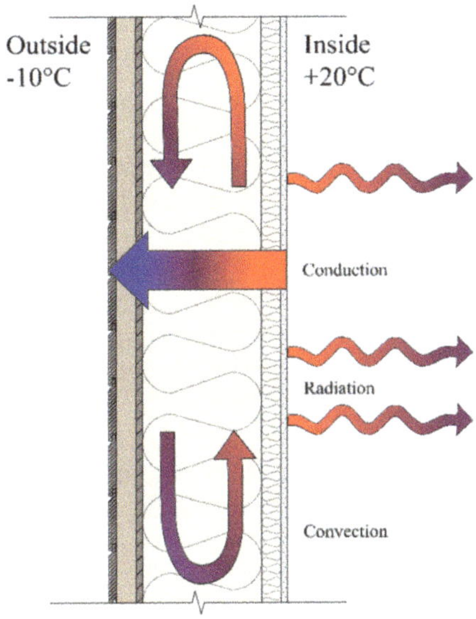

Figure 2.1. Heat transfer principles in a typical timber frame wall construction.

$$dQ = -\lambda \cdot \frac{\delta T}{\delta x} \cdot dA \cdot dt \text{ [Ws]} \tag{2.1}$$

$$q_l = \frac{\lambda}{d} \cdot (T_{\text{interior}} - T_{\text{exterior}}) \text{ [W m}^{-2}\text{]} \tag{2.2}$$

From Fourier's law, it is possible to calculate the thermal resistance, R, of a homogeneous material layer with thickness d, an area of 1 m^2 and a temperature difference of 1 K. For non-homogeneous components, the thermal resistance is calculated for each homogenous layer and summed to the total resistance for the component.

$$R = d/\lambda \text{ [m}^2\text{ K W}^{-1}\text{]} \tag{2.3}$$

The factor λ is the thermal conductivity of the material in W m^{-1} K^{-1}. Materials with high thermal conductivity will quickly transfer heat, while materials with low thermal conductivity are good insulators. The thermal conductivity of some conventional (building) materials is given in table 2.1. More values can be found in ISO 10456:2007 [3]. The thermal conductivity of porous materials increases when they are wet or when the humidity increases, which should be taken into account when calculating the thermal resistance. The higher conductivity of water explains this increase in thermal conductivity compared to air.

2.1.2 Heat convection

Convection is the heat transfer between a fluid and a surface material when a fluid with a temperature difference moves past the surface. Convection can be forced

Table 2.1. Thermal conductivities of common materials (data from [3]).

Material	Thermal conductivity [W m^{-1} K^{-1}]	Material	Thermal conductivity [W m^{-1} K^{-1}]
Air	0.025	Concrete, reinforced	2.30
Water, at 10 °C	0.60	Brick	1.30
Ice, at −10 °C	2.30	Glass	1.00
Snow	0.05–0.60	Soft wood	0.13
Steel	50	Hard wood	0.18
Aluminium alloys	160	Insulation	0.025–0.050

when it is caused by wind or mechanical devices, such as fans or pumps, or it can be naturally caused by buoyancy. Buoyancy is an upward force caused by the difference in density of the fluid. Fluids with a lower density will rise, and this also happens when part of a fluid is heated.

Convection is a complex mechanism, but the calculation of the density of heat flow rate can be simplified as in equation (2.4). The factor $h_{.c.}$ is the convective heat transfer coefficient in W m^{-2} K^{-1}. The heat transferred by convection is influenced by the type of fluid, the state of the fluid, the temperature difference between the fluid and the surface and the flow type (laminar or turbulent) [1, 2].

$$q_c = h_c \cdot (T_{\text{surface}} - T_{\text{fluid}}) \text{ [W m}^{-2}\text{]} \tag{2.4}$$

Only convection of air and convection in windows is significant for the heat transfer in buildings. Though air has a low thermal conductivity, convection causes heat losses in air pockets large enough for the air to move. However, when an air pocket with stationary air is created, the air is a satisfactory insulator. For closed and unventilated spaces filled with a gas, the convective heat transfer is calculated according to equation (2.5). In this equation, d is the width of the pocket and λ_c is the convection coefficient, which depends on the fluid characteristics, the size and orientation of the pocket and the temperature difference across the pocket.

$$q_c = \frac{\lambda_c}{d} \cdot (T_2 - T_1) \text{ [W m}^{-2}\text{]} \tag{2.5}$$

2.1.3 Heat radiation

Heat can also be transferred through electromagnetic waves. Radiant heat does not need a medium to be transferred in, which means it can be transferred through a vacuum. The energy depends on the wavelength, which is influenced by the temperature of the object. Warm objects emit short waves with high energy. The density of heat flow rate depends on the colour of the object described as the emissivity, and the surface temperature. The emissivity represents the percentage of incoming heat that is emitted, where 1 means the material is a perfect radiator and

0 means it is a perfect reflector. The density of radiant heat transfer, q_r, is a function of the emissivity, ε, the temperature difference between the radiating surface and its surroundings and the Stefan–Boltzmann's constant, σ. Equation (2.6) shows that equation (2.7) can be simplified into a linear form, where a large part of the equation is expressed as the radiative heat transfer coefficient, h_r [1, 2]. This coefficient is highly dependent on the temperature difference between the surface and the surroundings.

$$q_r = \varepsilon \cdot \sigma \cdot \left(T_{\text{surface}}^4 - T_{\text{surrounding}}^4\right) \text{ [W m}^{-2}\text{]} \tag{2.6}$$

$$q_r = h_r \cdot (T_{\text{surface}} - T_{\text{surroundings}}) \text{ [W m}^{-2}\text{]} \tag{2.7}$$

2.1.4 Heat transfer through envelope components

A component consisting of parallel homogenous layers has an equal density of heat flow in all layers when the temperature difference does not change. The heat transfer is influenced by the thermal resistance, R, of the component. This indicates the insulating properties of the material; a high thermal resistance means the material insulates well. The thermal resistance of the component is equal to the sum of thermal resistance from each layer and the transmission resistance of the internal and external surfaces. The surface transmission resistance depends on the heat flow direction. The exterior surface transmission resistance is equal to 0.04 m^2 K W^{-1}, and the interior surface transmission resistance is equal to 0.10, 0.13 and 0.17 for upward, horizontal and downward heat flows, respectively [1, 2].

$$q = \frac{1}{(R_{si} + R_1 + \ldots + R_n + R_{se})}(T_i - T_e) \text{ [m}^2\text{ K W}^{-1}\text{]} \tag{2.8}$$

The U-value is a commonly used term to express how much heat (W) is transferred through 1 m^2 component with a temperature difference of 1 K between outside and inside. A high U-value means that the component lets through much of the heat and a low U-value means the component insulates well. The U-value is the inverse value of the total thermal resistance, R_{tot}, as given in equation (2.9). Sometimes a correction factor, ΔU, is added to take into account air slits, anchors and braces in the insulation layer.

$$U = \frac{1}{R_{tot}} = \frac{1}{(R_{si} + R_1 + \ldots + R_n + R_{se})} \text{ [W m}^{-2}\text{ K]} \tag{2.9}$$

In reality, the U-value can differ significantly from the calculations, either in a positive or negative direction. There is uncertainty in the calculation because standardized values are used. For example, the exterior and interior surface resistance may differ from standard values, and the thermal conductivity of porous materials is influenced by moisture. Both of these can be less or more than expected. Poor handling of the building materials on-site and improper construction with gaps and leaks can result in significantly worse U-values and increase the number of thermal bridges. Conversely, with smart solutions of construction joints and proper execution on the building site, the thermal properties of the envelope can become better than estimated.

2.1.4.1 Homogeneous components

As the heat transfer through a component is equal in each of its layers, we can assume that conduction and the temperature distribution transfer of the heat is one-dimensional and stationary [2]. With these assumptions, the total thermal resistance of a component that consists of only homogeneous layers is calculated by simply adding the thermal resistance of all the layers together. The thermal resistance of a material layer is calculated according to equation (2.8). The U-value of the component is then calculated, as shown in equation (2.9). The thermal resistance of an air layer in the component depends on the airflow in the layer. If the space is not ventilated, the thermal resistance is calculated as normal. If the space is slightly ventilated or well ventilated, the thermal resistance is 50% and 0%, respectively, of the thermal resistance of a non-ventilated air layer. In the case of a well-ventilated space, the thermal resistance of all additional layers (e.g. cladding) towards the exterior is discarded; however, in this case, the exterior surface transmission resistance can be assumed 0.13 m^2 K W^{-1} for walls and 0.10 m^2 K W^{-1} for roofs instead of 0.04 m^2 K W^{-1}. As insulation has lots of small air pockets, the insulating value of insulation materials with open pores, such as mineral wool, is highly influenced by the airflow. Therefore, it is important that the insulation is placed behind a wind barrier to prevent heat transfer through convection [1].

2.1.4.2 Non-homogeneous components

Not all components consist of only homogenous layers. Components that have non-homogenous layers, such as timber frame walls, do not have the same U-value throughout the whole structure. The U-value for the whole component is the arithmetic mean of the upper and lower limit of thermal resistance. When the conductivity difference between the materials in the non-homogenous layers is not too significant, this method is a good indication of the overall R-value. However, when the ratio between the upper and lower limit is higher than 1.5, the method will give a wrong estimate because it indicates that there are one or several significant thermal bridges.

In the upper limit, it is assumed that the heat transfers only in one dimension through the layers, perpendicular to the surface. The upper limit equation (2.11) is therefore calculated as the average total thermal resistance for all sections, corrected for the relative area of each section (f in %). For example, in a timber frame wall, there are two sections: one through the insulation layer between the studs and one through the studs (figure 2.2). In this case, f_a is the percentage of wooden studs in the wall, $R_{tot,a}$ is the thermal resistance of the section through the studs, f_b is the percentage of insulation and $R_{tot,b}$ is the thermal resistance of the section

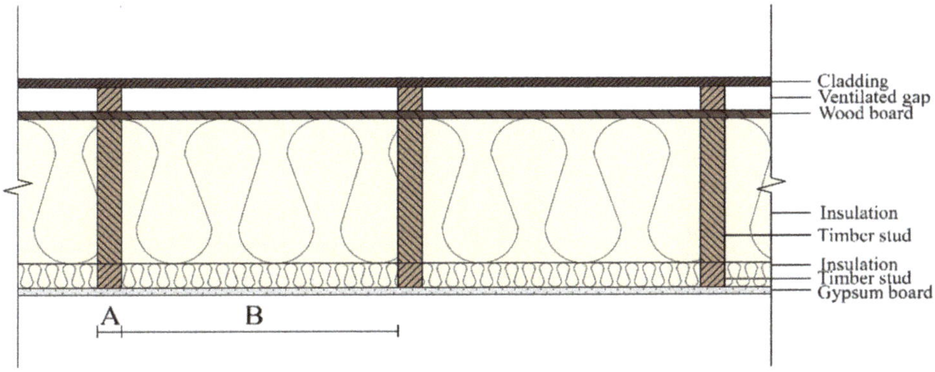

Figure 2.2. Defining sections for calculating the *U*-value of a component with non-homogenous layers, such as a timber frame wall.

through the insulation. The percentage of wooden studs is often assumed to be 12% or 9% when the studs are c/c 600 mm and have a width of 48 mm or 36 mm, respectively. However, the percentage increases when there are many door and window frames and should be calculated for the specific case to get accurate results. For the lower limit equation (2.12), it is assumed that the layers are isothermal (i.e. the temperature varies in one dimension only). The thermal resistance is the summation of the thermal resistance of each layer where an alloy thermal resistance is calculated for the non-homogeneous layers. For the timber frame wall example, f_a is the percentage of wooden studs in the wall, R_{aj} is the thermal resistance of the non-homogenous layer through the studs, f_b is the percentage of insulation, and R_{bj} is the thermal resistance of the non-homogenous layer through the insulation. The thermal resistance of the ventilated gap behind the cladding and the cladding itself is excluded from the calculation. Instead, the thermal resistance is taken into account in the exterior surface transmission resistance, as mentioned above.

$$R_{\text{tot}} = \frac{R_{\text{upper}} + R_{\text{lower}}}{2} \ [\text{m}^2\ \text{K W}^{-1}] \tag{2.10}$$

$$R_{\text{upper}} = \frac{1}{\frac{f_a}{R_{tot,a}} + \frac{f_b}{R_{tot,b}} + \ldots + \frac{f_n}{R_{tot,n}}} \ [\text{m}^2\ \text{K W}^{-1}] \tag{2.11}$$

$$R_{\text{lower}} = R_{si} + R_1 + \frac{1}{\frac{f_a}{R_{aj}} + \frac{f_b}{R_{bj}} + \ldots + \frac{f_n}{R_{nj}}} + \ldots + R_n + R_{se} \ [\text{m}^2\ \text{K W}^{-1}] \tag{2.12}$$

2.1.4.3 Windows

The *U*-value for windows is often given as a merged value for the glass panes, the frame and thermal bridges. Window producers often provide the total *U*-value for the window. This U-value counts for the whole area of the window, including the frame. In equation (2.13), A_g and A_f are the areas of the glazing and frame, respectively, ψ is the linear thermal bridge value, and l_g is the length of the thermal bridges. The spacers between the windowpanes and the joints between the glazing and the frame are thermal bridges in the component and can significantly influence the total U-value of the window.

$$U_w = \frac{A_g \cdot U_g + A_f \cdot U_f + l_g \cdot \Psi}{A_g + A_f} = [\mathrm{W\,m^{-2}\,K}] \tag{2.13}$$

The U-value for the glass itself is calculated as for homogenous components. The U-value of the windowpane is influenced by the number of panes and the gaps between the panes. The thermal resistance of the gap depends on the gas inside the gap, temperature difference, the mean temperature and the emissivity. A lower emissivity of the glass panes facing a gap will reduce the heat transfer by radiation, and the U-value improves. This can be achieved by applying coatings to the windowpanes, such as a thin metal coating or an oxide coating. These coatings reflect the heat but are thin enough so that the view is not obstructed, and daylight can penetrate. The U-value can also be improved by using gas with a lower thermal conductivity than air, such as argon or krypton. The gap must be sealed as puncturing will result in the gas escaping. The window frame U-value depends on the material used, its size, and how the glass is integrated into the frame. Window frames are often made of wood, plastic or of metal casing, such as aluminium, with a thermal break in the frame profile.

2.1.4.4 Ground

Heat transfer through components that border the ground is different from those that face air because of the thermal properties of soil. The heat transfer varies over time as the ground temperature fluctuates with the outdoor weather conditions, and it varies across the surface area of the basement wall or floor. The heat flow is reduced the farther the component is from the outer limit of the building and the deeper it is underground. Because of this, the previously shown method that assumes one-dimensional heat flow under stable conditions over the surface area is not sufficient. The method to calculate the correct U-value for structures bordering soil is described in ISO 13370 [4].

2.1.5 Thermal bridges

Thermal bridges are sections in the envelope that have a significantly higher U-value than the surrounding parts of the building envelope. This can occur when materials with a higher conductivity break through the insulating layer. Also, this can occur when there is a change in insulation thickness either because of the construction or through poor execution. Thermal bridges also occur in places where the surface area towards the exterior is larger than the surface area towards the interior, such as in corners. In inverse corners (i.e. larger interior surface), this results in positive thermal bridges. Thermal bridges should be avoided as they cause several problems, both for the building construction as for the users inside. As the U-value of a thermal bridge is higher than its surroundings, the heat flow through this part of the construction is increased. If many or large thermal bridges occur, such as a concrete balcony without a thermal break or curtain walls with metal frames, this can have a significant effect on the space heating demand and energy bill. Because of the increased heat losses, the local surface temperature is lower than its surroundings. This can cause discomfort for the users because of asymmetrical heat transfer in the room, which can cause drafts, cold corners and cold floors. The temperature difference can also cause damage to building construction. If the humidity in the room is high, such as in a kitchen or bathroom, condensation can occur at cold surfaces, such as windows and thermal bridges. If the surface material absorbs moisture, its thermal conductivity will increase, and the total thermal resistance decreases further. Condensation can also cause moisture damage and mould growth in the construction and on interior surfaces.

The thermal bridge value takes into account extra heat losses caused by the thermal bridge. This value is additional to the heat losses through the envelope without thermal bridges. There are two types of thermal bridges: linear (ψ) and point (χ). Linear thermal bridges are the most common and occur over a certain length, such as a thermal bridge between the window frame and wall or between the balcony and floor. Point thermal bridges, such as anchors through the insulation, are generally neglected unless they are significant. The linear and point thermal bridge values should be calculated for each type of thermal bridge, but sometimes a normalized value for linear thermal bridges is given. This value is a typical value for all thermal bridges in the building, for example, for a typical concrete building built in the 1980s.

2.1.6 Thermal mass

Thermal mass is a term used to describe building materials that can store and release energy in the form of heat. Thermal mass is equivalent to the heat capacity and is expressed as the amount of heat (J) required to heat up the material by 1 °C. Because of its ability to store heat, thermal mass causes a delay of temperature fluctuations and a reduction of the amplitude of the fluctuations, as illustrated in figure 2.3. In other words, the indoor temperature becomes more stable throughout the day and responds more slowly to outdoor temperature fluctuations. During the day, solar heat is absorbed and stored in the exposed thermal mass. As it takes time to heat up the material, this prevents overheating in the room during the daytime. It reduces the peak cooling load and improves indoor comfort. When the temperature in the room drops, the heat is released from the thermal mass and provides a more comfortable night-time temperature in the winter. When the heat is unwanted, night ventilation can be used to flush out the excess heat.

In buildings, we are interested in materials that have a high heat capacity as well as a high density to maximize the heat storage and minimize the volume. In this case, thermal mass is often compared by referring to the specific heat of materials. Specific heat is the heat capacity per weight, in J K^{-1} kg^{-1}. To quickly assess the space

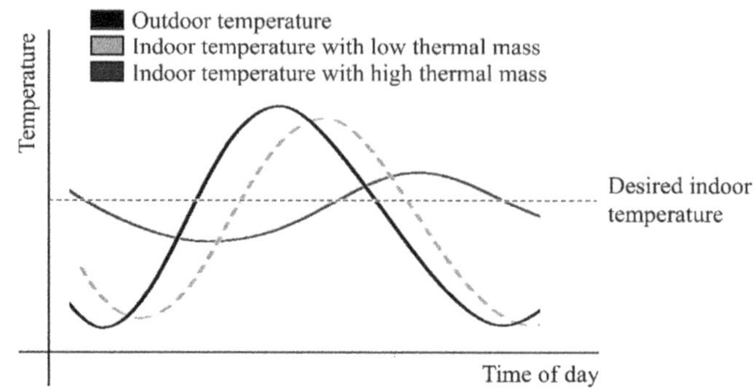

Figure 2.3. Effect of thermal mass on the indoor temperature as a response to the outdoor temperature when no heating or cooling systems are active.

Table 2.2. Thermal properties of common interior building materials (date from [5]).

Building material	Density [kg m^{-3}]	Specific heat capacity [J kg^{-1} K^{-1}]	Thermal conductivity [W m^{-1}·K^{-1}]	Volumetric heat capacity [kJ m^{-3}·K^{-1}]	Thermal mass efficiency
Air	1.23	1008	0.025	1.24	None
Carpet	200	1300	0.06	260	None
Steel	7800	450	50	3510	Low
Soft wood	500	1600	0.13	800	Low
Hard wood	700	1600	0.18	1120	Low
Gypsum board	900	1050	0.25	945	Low
MDF	600	1700	0.14	1020	Low
Linoleum	1200	1400	0.17	1680	Low
Ceramic tiles	2300	840	1.30	1932	Medium
Brick	1750	1000	1.30	1750	Medium
Reinforced concrete	2300	1000	2.30	2300	High
Concrete	18002400	10001000	1.152.00	18002400	Medium/High
Water	1000 (10 °C) 990 (40 °C)	41874187	0.60 (10 °C) 0.63 (40 °C)	41874145	Very high

efficiency and thermal mass potential of a specific building material, specific heat can be multiplied with the density to get the heat capacity per volume. The higher the volumetric heat capacity, the more heat can be stored in a space-efficient way.

Table 2.2 lists some properties of conventional building materials and their effectiveness as thermal mass. In general, it can be said that heavier materials are more suitable as thermal mass. An exemption is water, which has the property that it can store a significant amount of heat in comparison to its weight and volume; however, water is less easy to incorporate into the building interior. In addition to specific heat and density, the thermal conductivity is essential too. If the thermal conductivity is too low, then the material will function as an insulator instead of heat storage. A high thermal conductivity will result in heat passing through the material rather than being stored. Because of this, metal products are not suitable as thermal mass even though their heat capacity per volume is high. Floor coverings such as ceramic tiles have a high heat capacity per volume but have a limited effect due to their slimness. These are best used in combination with a high thermal mass floor, such as concrete. Floor and wall coverings insulate the underlying material and can reduce the heat storage effect of the underlying material, depending on the thermal conductivity of the covering. The material should be exposed to achieve the largest thermal mass effect.

2.2 Energy balance of the building

The energy balance of a building consists of energy gains, losses and storage from different sources. It can be used to calculate the energy consumption of a building. There are three methods to calculate the energy consumption of a building: empirical, static and dynamic. Empirical methods use field measurements as a foundation for the calculation. These are very simplified methods and require several correction factors due to their inaccuracy. However, empirical methods give a quick and rough estimate of the total energy consumption of a building. Static methods are based on mathematic models and assume that the conditions in the building are static during the calculation period. This method is more detailed and accurate than the empirical method but still requires correction factors to take into account dynamic conditions. Dynamic methods are the most accurate and comprehensive. These methods divide the total simulation period into time steps, often one minute or hour, in which the conditions are assumed to be static. After every time step, the conditions for each zone are re-evaluated and adjusted when changes occur. For example, a cloud blocking the Sun or a change in the number of occupants. This results in a precise outcome but can be extremely time-consuming to perform. A complex calculation like this is executed by simulation programs, though the other two methods can be done without programs [2].

The total energy consumption consists of energy for lighting and appliances, energy for domestic hot water (DHW) and energy for HVAC systems. The energy use for lighting, appliances and DHW depend mostly on the number of occupants, the building function and the consumption pattern of the users. In cold climates, the energy for the HVAC system, especially heating, typically dominates in the energy balance. However, in low or zero energy buildings the share of lighting, appliances and DHW becomes significantly more important [5].

$$Q_{tot} = Q_{function} + Q_{DHW} + Q_{HVAC} \text{ [kWh year}^{-1}\text{]} \tag{2.14}$$

where Q_{tot} is the total energy consumption, $Q_{function}$ is the energy consumption for lighting and appliances, Q_{DHW} is the energy used for DHW and Q_{HVAC} is the energy used for HVAC systems.

2.2.1 Energy use for lighting and appliances

The energy consumption for lighting and appliances is mostly electrical. The final energy consumption of lighting and appliances depends on the number of occupants, the use pattern and the typical energy use for the device. The energy use for lighting can be calculated using equation (2.14) [6]. The total energy use, W_t, for lighting during a period, t, can be calculated using the following equations:

$$W_t = W_{L,t} + W_{p,t} \text{ [kWh t}_s^{-1}\text{]} \tag{2.15}$$

$$W_{L,t} = \sum((P_n \cdot F_c) \cdot F_o((T_D \cdot F_D) + t_N))/1000 \text{ [kWh t}_s^{-1}\text{]} \tag{2.16}$$

$$W_{P,t} = \sum((P_{pc} \cdot t_s) + (P_{em} \cdot t_e))/1000 \text{ [kWh t}_s^{-1}\text{]} \tag{2.17}$$

where $W_{L,t}$ is the estimated energy for lighting and $W_{P,t}$ is the energy used in the standby modus. P_n is the luminaire power when the lighting is operating at maximum power, F_c is the constant illuminance dependency factor, F_o is the occupancy dependency factor, T_D is the daylight time, F_D is the daylight dependency factor, and t_N is the daylight absence time. P_{pc} is the standby power when lighting is not operated, t_s is the chosen time step for the number of standby hours, P_{em} is the power to charge batteries (emergency power) and t_e is the battery charge time. A more detailed explanation of these variables and how to calculate them is given in EN 15193–1 [6]. From these formulas, the annual energy use per m^2, *LENI*, for lighting can be calculated using equation (2.20).

$$ENI = \sum\left(\frac{8760}{t_s} \cdot W_t\right) \Big/ A \text{ [kWh m}^{-2}\text{ year]} \tag{2.18}$$

As this method is quite complicated, and many input parameters are needed to use this comprehensive method, a simplified method can be used to estimate the annual energy use from lighting. The lighting energy consumption for a zone is:

$$LENI_{sub} = \left(F_c \cdot \left(\frac{Pj}{1000}\right) \cdot F_o((t_D \cdot F_D) + t_N)\right) + 1.0 + 1.5 \text{ [kWh m}^{-2}\text{ year]} \tag{2.19}$$

In this formula, the standby energy consumption P_{pc} and P_{em} are assumed as 1.0 and 1.5, respectively. P_j is the power density for the area in W m^{-2}. For the other variables, national or default values can be used. When there are different lighting zones in the building, the energy consumption for each zone, i, should be calculated and then converted to the average annual consumption per m^2 using the following formula:

$$LENI = \frac{\sum_{i=1}^{n}(LENI_{sub,i} \cdot A_i)}{A} \text{ [kWh m}^{-2}\text{ year]} \tag{2.20}$$

The total annual energy use of appliances can be estimated by:

$$Q_{A,t} = \sum((P_A \cdot t_{on}) + (P_{sb} \cdot t_{sb}))/1000 \text{ [kWh year}^{-1}\text{]} \tag{2.21}$$

where $Q_{A,t}$ is the total energy use of appliances, P_A is the power of the appliance when it is switched on in Watt, t_{on} is the number of annual hours the appliance is switched on, P_{sb} is the power the appliance consumes in standby mode and t_{sb} is the number of annual hours the appliance is in standby mode. For this estimation, the energy use of each appliance in the house is calculated and then summed. The equation does not take into account the behaviour of occupants.

2.2.2 Energy use for domestic hot water

The energy use for DHW depends on the efficiency and losses of the system, the water temperature and the volume of consumed water. The volume is dependent on the number of users, the function of DHW (e.g. hygienic or recreational function) and user patterns and habits. Because of this, national standards often give average values for their country in L/person, L/day or kWh m^{-2} year. It is therefore recommended to use these values when calculating the total energy use for DHW.

2.2.3 Energy use for HVAC systems

The final and most complicated factor in the energy balance is the energy use for HVAC systems. In cold climates, the largest share of energy use is for space heating. Active cooling systems are not always present and needed, though in buildings with high internal gains (such as offices) there may appear a cooling demand. Since space heating dominates in cold climates, most of this section will focus on how to calculate this. The energy consumption for each load is calculated by multiplying the demand by the system efficiency. The system efficiency is influenced by the type and age of the heating system and the energy source. The space heating demand is influenced by the design of the building, the use of the building and outdoor conditions. Table 2.3 summarizes the parameters that can affect the space heating demand of a building.

Before the space heating demand can be calculated, the building has to be divided into thermal zones. A thermal zone is an 'internal environment with assumed sufficiently uniform thermal conditions to enable a thermal balance calculation' [4]. Adjacent zones can be merged when they are of the same space category or function, have large openings between them or have similar thermal use conditions. Adjacent zones that are thermally unconditioned can also be combined. Merging zones reduces the calculation time, but if not done carefully will result in errors.

2.2.4 Heating demand

The heat balance for a zone, i, with zone volume is V_i, for a time step, ts, is given by equations (2.22) and (2.23). To decrease the net energy demand for space heating, either the heat losses should be decreased, or the heat gains increased. In cold climates, most energy savings can be achieved by decreasing heat losses. Increasing

Table 2.3. Parameters that influence space heating energy demand [1, 2, 5].

	Variable	Influenced by	Effect
Outdoor conditions	Temperature	Weather and local climate	A larger temperature difference between outdoor and indoor results in a higher space heating demand.
	Wind speed		A higher wind speed or less sheltering increases infiltration heat losses if the outdoor temperature is lower than the indoor temperature.
	Solar radiation		More solar radiation reduces the space heating demand because of solar heat gains but can also increase the cooling demand.
Building design	Thermal properties	*U*-value	A lower *U*-value results in decreased heat transmittance through the envelope and a lower space heating demand.
	Compactness	Shape and form	The demand for space heating decreases when the building is more compact, because there is less heat loss (smaller surface area).
	Window area	Orientation, area, *U*-value	Windows influence heat losses because of their thermal properties which are generally worse than opaque envelope (larger area is more heat losses) and influence heat gains by letting through solar radiation (larger area is more heat gains). In winter, the heat losses are often higher than the solar heat gains.
	Infiltration	Airtightness	A more airtight building will have less infiltration heat loss when the outdoor temperature is lower than the indoor temperature.
	Thermal mass	Thermal capacity	The space heating demand can be reduced if the building has a high thermal capacity, is well insulated and the heating set-point is constant. In other cases, there are no positive effects.

(*Continued*)

Table 2.3. (*Continued*)

	Variable	Influenced by	Effect
Building use	Indoor temperature	User behaviour, control strategies	A higher set-point for indoor temperature will increase the space heating demand. Temperature setback can decrease the space heating demand.
	Ventilation	Control strategies, heat recovery	Lower ventilation rates during the heating season will result in a lower space heating demand. Heat recovery also lowers the space heating demand.
	Internal gains	People, equipment and lighting	More internal gains will lower the space heating demand but can also result in a higher cooling demand.
	DHW	System efficiency and losses	Draining hot water results in heat losses. There are often heat losses in the system as well, but when these are transferred to the heated area, they are heat gains.

the heat gains can result in overheating and a need for active cooling, which increases the total energy consumption of a building and decreases the indoor comfort level. Therefore, one should be careful with increasing heat gains.

$$Q_{tot,heat,i,ts} + Q_{tot,loss} + \eta_{use} \cdot Q_{tot,gain} = 0 \text{ [kWh]} \tag{2.22}$$

$$Q_{tot,heat,i,ts} = [(\Phi_{T,i,ts} + \Phi_{V,i,ts}) - \eta_{use,i,ts}(\Phi_{sol,i,ts} + \Phi_{int,i,ts})]\Delta t_{ts} \text{ [kWh]} \tag{2.23}$$

where $Q_{tot,heat,i,ts}$ is the net energy demand for space heating, most often expressed in kWh and sometimes in Joules. $\Phi_{T,i,ts}$ is the transmission flow to the surroundings in Watt, $\Phi_{V,i,ts}$ are the ventilation and infiltration heat flows in Watt. The transmission and ventilation flow, $Q_{tot,loss}$, are negative (heat losses) in cold climates; therefore, they are positive on the other side of equation (2.22). $\eta_{use,i,ts}$ is the utilization efficiency of heat gains, $\Phi_{sol,i,ts}$ are the solar heat gains in Watt, $\Phi_{int,i,ts}$ are the internal heat gains in Watt and Δts is the chosen time step in seconds. When calculating by hand, the time step is often a month. Simulation programs often use one hour or minute as a time step. The annual heating demand follows by adding up all time steps, whether it be months, days or hours. The delivered energy is calculated by dividing the heating demand by the efficiency of the system. If the system efficiency varies throughout the year, this should be done in time steps. If several heating systems are used, the delivered energy should be calculated per energy source and then added together [2, 5, 7].

$$Q_{heat,delivered} = \sum \left[\sum_{m=1}^{12} \frac{Q_{heat,month}}{\eta_{month}} \right] [kWh] \tag{2.24}$$

Heat losses occur when there is a temperature difference between the interior and exterior of a building component. The total heat losses consist of transmission heat losses through the building envelope and ventilation heat losses through infiltration and the ventilation system.

2.2.4.1 Transmission heat losses

Transmission losses are influenced by the difference in indoor and outdoor temperature, the surface area of the component towards the exterior and the thermal transmittance of the material. This means that transmission losses can be reduced by decreasing the temperature difference, decreasing the component area or by decreasing the thermal transmittance of the component. The latter is straightforward to achieve in practice. The temperature difference can be decreased by lowering the set point for indoor temperature, although it should not negatively affect the thermal comfort or the construction. The transmission flows to the surroundings are flows to the exterior, to other zones with different temperatures and to zones with the same indoor temperature. The latter type of flows are ignored because there is no heat transfer between zones with the same temperature. The transmission flows are calculated by multiplying the heat transfer coefficients between zones with the temperature difference between zones (see equation (2.25)). H_T, i.e. is the heat transfer coefficient between the zone and outside and $H_{T,ij}$ is the heat transfer coefficient between the zone and another zone, j. The heat transfer coefficient is calculated according to equation (2.26). When calculating the heat transfer coefficient between the zone and other interior zones, the factor for heat transfer to the ground is removed from the equation.

$$\Phi_{\mathrm{T,i}} = H_{T,ie}(\theta_i - \theta_e) + \sum_{j=1}^{m} H_{T,ij}(\theta_i - \theta_j)\ [W] \tag{2.25}$$

$$\begin{aligned} H_T = &\sum_{opaque} U_{op} \cdot A_{op} + \sum_{window} U_w \cdot A_w + \sum_{ground} a_{fl} \cdot U_{fl} \cdot A_{fl} \\ &+ \sum_{linear} \psi \cdot L + \sum_{local} \mathrm{X}\ [\mathrm{W/K}] \end{aligned} \tag{2.26}$$

where θ_e is the exterior temperature, θ_i is the temperature in the zone and θ_j is the temperature in the adjacent zone. The symbol U refers to the thermal transmittance (U-value) of the component in W m^{-2} K^{-1}, A is the surface area in m^2, ψ is the thermal transmittance of linear thermal bridges in W m^{-1}K^{-1}, L is the length of the linear thermal bridges in m and χ is the thermal transmittance of local or point thermal bridges in W m^{-2} K^{-1} [2, 7].

2.2.4.2 Ventilation heat losses

The second part of the total heat losses is heat transfer through ventilation, $\Phi_{V,j}$. This flow consists of two elements: heat transfer through mechanical ventilation and heat transfer through infiltration. Infiltration is unwanted air leakage through the building envelope and openings. This is influenced by the airtightness of the building, which depends on the properties of the airtight layer and the execution. Generally, older buildings are leaky and have a much higher infiltration rate than new buildings. Infiltration is wind-driven and thus influenced by the local climate. However, if a mechanical ventilation system is installed, the exhaust airflow rate will decide the infiltration rate. If the ventilation system creates negative pressure, the air will be supplied through the leaks in the building envelope, and this can significantly increase the heating demand of the building. The ventilation system influences the heat losses through ventilation. The heat losses increase proportionally with the air volume and with the temperature difference between interior and exterior. The heat that is removed with the exhaust air can have a significant impact on the total heat losses and thus the energy consumption. This can be improved by installing a heat recovery system that transfers the heat from the exhaust air to the supply air to provide free heating. Industry buildings often recover waste heat from the ventilation system and chimneys to provide free space heating for themselves and sometimes surrounding buildings. The infiltration rate has an impact on the total heat recovered, as this type of supply air does not pass through the ventilation system.

Equations (2.27) to (2.30) show the calculation for heat losses through ventilation. If the building does not have a mechanical ventilation system, only the heat transfer through infiltration is considered. The air changes in the mechanical ventilation rates due to infiltration are taken into account when calculating the heat transfer coefficient for infiltration [8].

$$\Phi_{\mathrm{V,i}} = (H_{V,i} + H_{inf})(\theta_i - \theta_e)\ [W] \tag{2.27}$$

$$H_{inf} = \rho_a \cdot c_a \cdot n_{inf} \cdot V \tag{2.28}$$

$$n_{inf} = \frac{n_{50} \cdot e}{1 + \frac{f}{e} \cdot \left(\frac{q_{sup} - q_{exh}}{V \cdot n_{50}}\right)} \tag{2.29}$$

$$H_{v,i} = \rho_a \cdot c_a \cdot (1 - \eta_{rec}) \cdot \sum_k b \cdot q_{vent} \cdot f_{vent} \tag{2.30}$$

The term $\rho_a{\cdot}c_a$ is the heat capacity of air per volume and is 0.33 Wh m^{-3} K, n_{inf} is the infiltration rate in h^{-1}, V is the air volume of the zone, n_{50} is the infiltration rate at 50 Pa, e and f are factors for terrain sheltering and q_{sup} and q_{exh} are the air flow rates for supply and return air in m^3 h^{-1}, respectively. The terms b and f_{vent} are correction factors as described in ISO 52016–1 [9], q_{vent} is the airflow rate in m^3 h^{-1} and η_{rec} is the heat recovery efficiency between 0 and 1. If there is no heat recovery, η_{rec} is equal to zero. It should also be noted that the equations for calculating the ventilation losses do not take into account the opening of windows or doors. This will cause extra heat losses, especially on cold days.

2.2.4.3 Intermittent heating and reheating capacity

The equations for transmission and ventilation losses have a constant value for the temperature in the zone. When intermittent heating is installed, the total heat losses in kWh can be calculated with equation (2.31). In this equation, N_j is the number of sub-periods within the time step, $\theta_{iad,j}$ is the indoor adjusted temperature in the sub-periods, t_j is the length of the sub-period and $H_{tot,j}$ is the total heat losses of transmission and ventilation in the sub-period [2].

$$\mathrm{Q_{T,i}} = \sum_{j=1}^{N} N_j \cdot H_{tot,j}(\theta_{iad,j} - \theta_e)t_j\ [kWh] \tag{2.31}$$

Zones that have intermittent heating require a certain heating capacity to increase the zone temperature to the new set point within an acceptable time limit. This is called the reheating capacity. The reheating capacity is influenced by the thermal capacity of the zone, the length of the reheating period, the temperature difference between the zone and the desired temperature and the properties of the control system [2]. A simple method to calculate the additional heating power is given in equation (2.32).

$$\Phi_{RH,i} = A_i \cdot \phi_{RH}\ [W] \tag{2.32}$$

where A_i is the heated floor area of the zone and φ_{RH} is the reheating factor that depends on the length of the reheating period and the temperature setback. This factor can either be described based on the duration of the disuse period (in case the zone temperature drop is unknown) or based on the temperature setback when there are two defined set-points [7]. The factors based on the temperature setback are given in table 2.4. The setback temperature ($\Delta\theta_{sb}$) can also be estimated according to equation (2.33), where $\theta_{i,des}$ is the desired indoor temperature, $\theta_{e,sb}$ is the external temperature during the setback period, $t_{sb,i}$ is the duration of the setback period and τ is the building time constant, as calculated in equation (2.34). The total power demand for a zone is then calculated according to equation (2.34).

$$\Delta\theta_{sb} = (\theta_{i,des} - \theta_{e,sb}) \cdot \left(1 - e^{-\frac{t_{sb}}{\tau}}\right) [K] \tag{2.33}$$

$$\Phi_{tot,i} = \Phi_{T,i} + \Phi_{V,i} + \Phi_{RH,i} [W] \tag{2.34}$$

2.2.4.4 Domestic hot water losses

Hot water that is drained generally leaves the building with a higher temperature than when it entered. Heat losses through hot water can be significant if the building uses large amounts of water. In that case, it might be beneficial to install a heat recovery system for drainage water. The heat losses from hot water are often not included in calculations; however, in national standards may include a standardised or averaged value for the heat loss and energy use from hot water per building typology. This value can be used to supplement the calculation of the energy consumption [1].

2.2.4.5 Heat gains

By far the largest share of the heat gains is from the heating systems in the building. The energy consumption depends on the efficiency of the system, the flexibility of the system, the effect and reheating capacity and finally on the energy demand of the

Table 2.4. Reheating factor based on the setback temperature in a zone with an air change rate of 0.1 h^{-1} (data from [7]).

Setback temperature	1 K		2 K		3 K		4 K		5 K	
Thermal capacity	Light	Heavy	Light	Heavy	Light	Heavy	Light	Heavy	Light	Heavy
Reheating period	Reheating factor, φ_{RH}									
0.5 h	12	12	27	28	39	44	50	60	—	—
1 h	8	8	18	21	26	34	33	48	—	—
2 h	5	5	10	15	15	25	20	35	43	85
3 h	3	3	7	12	9	20	14	29	33	75
4 h	2	2	5	10	7	18	10	26	28	72

building. This factor is the unknown variable that is calculated through equation (2.35). This equation calculates the space heating demand, which can be converted to the energy consumption for the space heating system by multiplying with the system efficiency. Other heat gains can reduce the demand for space heating. The total free heat gains, $Q_{G,i}$, of a building consist of direct heat gains from solar energy, $Q_{sol,i}$, and internal sources, $Q_{int,i}$ such as people, lighting and appliances, and indirect heat gains that come from thermally unconditioned zones (ztu), such as a sunspace.

$$Q_{G,i} = Q_{sol,i} + Q_{int,i} \text{ [kWh]} \tag{2.35}$$

2.2.4.6 Solar heat gains

Even in cold climates, the Sun has a significant contribution to the total heat gains. The solar heat gains in zone i consist of three factors: direct solar gains through transparent components, direct solar gains through opaque building components and indirect solar heat from adjacent unconditioned zones, k [4, 9].

$$Q_{sol,i} = Q_{sol,dir,i} + \sum_{k=1}^{n}[(1 - b_{,k}) \cdot F_{i,,k} \cdot f_{red,k} \cdot Q_{sol,dir,k}] \, [kWh] \tag{2.36}$$

where $Q_{sol,dir,i}$ is the direct solar heat gain to the zone, $Q_{sol,dir,k}$ is the direct solar heat gain in an adjacent thermally unconditioned zone, b_k is an adjustment factor for the unconditioned zone, $F_{i,k}$ is the distribution factor for the gains attributed to zone i from the unconditioned zone and $f_{red,k}$ is a reduction factor to avoid overestimation of the gains from zone k.

The direct solar gains to the zone through each component, n, are calculated according to equations (2.37), (2.38) and (2.39). The solar heat gains through opaque elements and from adjacent zones normally are only a fraction of the total solar gains and are often neglected. Because the solar radiation varies throughout the year, solar gains should be calculated per month. The amount of solar heat that enters through a window is proportionally linked to the glazed area. The percentage of light that is let through is expressed as the solar factor, g. This factor depends on properties such as the type of glass, coatings and solar shading systems.

$$Q_{sol,dir,i} = \sum_{n=1} Q_{sol,dir,win,n} + \sum_{n=1} Q_{sol,dir,op,n} \text{ [kWh]} \tag{2.37}$$

$$Q_{sol,dir,win,n} = g_t \cdot A_w \cdot (1 - F_{frame}) \cdot F_{s,w} \cdot H_{irr} - Q_{sky,w} \text{ [kWh]} \tag{2.38}$$

$$Q_{sol,dir,op,n} = \alpha_{rad,o} \cdot R_{se,o} \cdot U_{tot,o} \cdot A_o \cdot F_{s,o} \cdot H_{irr} - Q_{sky,o} \text{ [kWh]} \tag{2.39}$$

where in equations (2.38) and (2.39), A is the component area in m^2, F_s is the reduction factor for shading or screening on the component, H_{irr} is the incoming solar irradiation in W m^{-2} and Q_{sky} is the extra heat flow due to radiation from the sky in kWh. g_t is the mean effective solar transmittance, F_{frame} is the frame area percentage of the window, $\alpha_{rad,o}$ is the absorption coefficient for the opaque component, $R_{se.o}$ is the external surface heat resistance of the component in m^2 W K^{-1}

and $U_{tot,o}$ is the thermal transmittance of the component in W m^{-2} K [4, 9]. The extra heat flow to the sky through radiation is calculated according to equation (2.40). In this formula, $F_{sky,n}$ is the view factor between the component, n, and the sky, h_{lr} is the long-wave radiative heat transfer coefficient in W m^{-2} K^{-1}, $\Delta\theta_{sky,air}$ is the difference between the sky temperature and the air temperature and Δts is the time step.

$$Q_{sky,n} = 0.001 \cdot F_{sky,\,n} \cdot R_{se,\,n} \cdot U_{tot,n} \cdot A_n \cdot h_{lw} \cdot \Delta\theta_{sky,air} \cdot \Delta ts\ [kWh] \quad (2.40)$$

2.2.4.7 Internal heat gains

The internal heat gains include all sources inside the building that emit heat to the zone as a side effect (i.e. it is not their primary function to heat up a building). The heat flow from internal gains consists of gains from people, gains from appliances, gains from lighting, gains from recoverable losses from DHW, gains from recoverable losses from HVAC systems and gains from recoverable losses from other processes. Gains from recoverable losses from HVAC systems, such as from fans or pumps, are not included in the calculation because they are integrated into the HVAC systems and do not transfer heat to the zone. To simplify the calculation of internal heat gains, often only people, appliances and lighting are considered (see equation (2.41)).

$$Q_{int,i} = Q_{people} + Q_{lights} + Q_{appliances}\ [kWh] \quad (2.41)$$

People emit heat through the surface of the skin and when breathing out. The emitted heat from one person depends on the level of activity and it can vary from 80 W (sleep) to 400 W (heavy work). When the room temperature increases, a larger share of the heat is emitted as latent heat due to evaporation of sweat. The number of people, level of activity, building function and occupancy schedules should be taken into account when calculating the heating and cooling load. The heat emitted from appliances and equipment varies greatly with the building type and use. Especially in households, the type of appliances follows market trends. In general, the number of appliances has increased, resulting in higher heat gains, and appliances are becoming more energy efficient, resulting in less heat emitted. Especially in industrial buildings, the heat emitted from machines and equipment can be so high that no additional heat is required. For a detailed graph of the internal gains from appliances, it is important to know the type of appliance, its location and the use hours. Some building functions require a minimum number of appliances regardless of the heated floor area, such as white goods in housing, which means the contribution of internal gains per m^2 can be significantly larger for a small apartment than for a large house. The heat gain of appliances is more evenly spread in non-residential buildings and the use pattern often equals work hours (though there are exceptions). Lighting can contribute significantly to the heat gains. The need for lighting varies with the seasons and follows the general trend of the heating demand (i.e. more lighting is used in the wintertime), when more heating is required. In commercial buildings, lighting can result in a cooling demand during the summer.

The heat gains from lighting depend on the type of lamps, the number of lamps and the use pattern. The use pattern is not only influenced by weather and daylight, but also by the electricity price. We see that in countries where electricity is cheap the users are less conscious about switching off lights [1, 2, 5].

One can imagine that mapping each individual internal gain element and its use pattern can be time-consuming, especially for bigger buildings. For most internal heat gains, national standardized values per building function exist, often given in kWh m^{-2}, and are multiplied with the heated floor area of the zone. This saves time on mapping the internal gains and use patterns as well as computational time when running simulations. However, actual values may differ significantly from standardized values due to user habits and local climate. In practice, internal heat gains are not evenly distributed throughout the day and seasons but are adjusted to the weather conditions and use patterns. Neither are they evenly distributed in a building as some rooms will have almost no internal heat gains (e.g. storage and hall) while other rooms have high internal heat gains (e.g. auditorium and kitchen). The standardized value is acceptable when the annual energy use for the whole building is evaluated. More detailed input values should be used for other energy and comfort evaluations, such as daily peak loads and overheating risk in individual rooms.

The heat gains from internal sources are often not so high compared to the other energy flows, especially in older buildings and domestic buildings. When buildings become more energy efficient, the heat losses will decrease significantly, and the internal gains will take up a bigger share of the total. In that case, the risk of overheating can increase during warm days when the heat losses through transmission and ventilation are low. This explains why for example office buildings often require cooling, even in cold climates. In contrast, these internal gains can significantly reduce the heating demand during the winter. Appliances and lighting do not only result in internal heat gains, they also add to the total energy consumption. For an energy efficient building we should therefore consider the internal gains so that the risk of overheating is decreased and the energy consumption for appliances and lighting is kept to a minimum.

2.2.4.8 Utilization factor

Not all heat gains can be used to heat up a zone. For example, on warm days some of the heat gains will be excessive and cannot be used to cover the heating demand. Therefore, the total heat gains are multiplied with a utilization factor, as can be seen in equation (2.44). The utilization factor depends on the building's time constant, the heat storage capacity and the gain and loss factor. The formula for calculating the utilization factor depends on the gain and loss factor, as can be seen in equations (2.42) and (2.43). In unheated zones or if the heating demand is zero, the utilization efficiency equals 1. The utilization factor is 0 when the outdoor temperature is equal to or higher than the indoor temperature set point [2, 7].

$$\eta = \frac{1 - \gamma^{a}}{1 - \gamma^{a+1}} \; if \; \gamma \neq 1 \tag{2.42}$$

$$\eta = \frac{a}{a+1} \; if \; \gamma = 1 \tag{2.43}$$

$$\gamma = \frac{Q_{gain}}{Q_{loss}} \tag{2.44}$$

$$a = a_0 + \frac{\tau}{\tau_0} \tag{2.45}$$

$$\tau = \frac{C}{(H_T + H_V)} \tag{2.46}$$

$$C = \sum \chi_j \cdot A_j = \sum \left[\sum_i \rho_{ij} \cdot c_{ij} \cdot d_{ij} \cdot Aj \right] \tag{2.47}$$

where γ is the gain and loss factor and a is a numerical parameter that depends on the time constant, τ. The values a_0 and τ_o are defined at a national level or can be found in tables. C is the heat storage of the zone and is calculated by summing the heat storage of each section, j, for a thickness of maximum 100 mm or until the first insulation layer. If the thermal capacity, χ, of each surface is known this can be multiplied with the area of the surface. If this is not known, the heat storage is calculated for each layer, i, in section, j, as in equation (2.47), where ρ_{ij} is the density, c_{ij} is the specific thermal capacity, d_{ij} is the thickness of the layer and A is the area of the section.

2.2.5 Cooling demand

The cooling demand of the building is calculated using the same elements that make up the space heating demand equation. Therefore, all parameters mentioned in table 2.5 also influence the cooling demand of the building. Internal gains and solar heat gains cause an increase of the cooling demand and heat losses through transmission cause a decrease. Heat losses through ventilation and infiltration can decrease and increase the cooling demand, depending on the difference between outdoor and indoor temperature because heat flows from the highest to the lowest temperature. If the outdoor temperature is higher than the indoor temperature (positive difference), the cooling demand increases. If the outdoor temperature is lower than the indoor temperature (negative difference), the cooling demand decreases, and the heating demand increases. Thermal mass can have a positive effect on the cooling demand, depending on the time delay and how the heat is flushed out. For example, if the time delay is 12 h and the heat is flushed out at night by natural ventilation, it will result in a lower cooling demand during the day. However, if the heat is not flushed out, the excess heat is released into the room and

Table 2.5. Parameters that influence the energy demand for space heating [2, 5].

	Variable	Influenced by	Effect
Outdoor conditions	Temperature	Weather	A high outdoor temperature, especially when higher than the desired indoor temperature, increases the cooling demand.
	Wind speed		A higher wind speed causes more infiltration. If the outdoor temperature is higher than the indoor temperature, it will increase the cooling demand.
	Solar radiation		More solar radiation increases the cooling demand because of solar heat gains.
Building design	Thermal properties	*U*-value	A well-insulated building retains heat better and can indirectly increase the cooling demand.
	Compactness	Shape and form	A compact building has less heat losses and can increase the cooling demand.
	Window area	Orientation, area, *U*-value	Large glazed areas result in high solar heat gains, which increases the cooling demand. Windows also contribute to heat losses which reduces the cooling demand.
	Infiltration	Airtightness	A more airtight building will have less infiltration heat losses and if the outdoor temperature is higher than the indoor temperature, infiltrated air will increase the cooling demand.
	Thermal mass	Thermal capacity	A high thermal mass can decrease the cooling demand, especially in combination with night-flush cooling.
Building use	Indoor temperature	User behaviour, control strategies	A higher set-point for indoor temperature will decrease the cooling demand.
	Ventilation	Control strategies, heat recovery	A higher ventilation rate will result in a decreased cooling demand if the outdoor temperature is lower than the indoor temperature. If the outdoor temperature is higher, a high rate will increase the cooling demand.
	Internal gains	People, equipment and lighting	More internal gains will increase the cooling demand.
	DHW	System efficiency and losses	This does not affect the cooling demand unless the system heat losses are transferred to the rooms.

the cooling demand is increased at the start of the next day. Other parameters that can influence the cooling demand of a building are listed in table 2.5 [2, 5].

The cooling demand is not often taken into account in cold climates as the space heating demand dominates the energy consumption; however, in well-insulated buildings and in buildings with high internal heat gain loads, a cooling demand can occur. This is often resolved by passive strategies, such as opening windows and using solar shading, but in some cases, this is not enough, and air-conditioning is necessary to maintain a comfortable indoor temperature. As mentioned, thermal mass in combination with night-flush ventilation is also effective to reduce the cooling load during the day.

2.2.6 Ventilation energy

The energy for ventilation consists of the energy for fans and the energy for heating or cooling the supply air. The energy consumption is influenced by the type of ventilation: natural, mechanical exhaust or balanced. A natural ventilation system does not require energy but will significantly increase the space heating demand. In addition, it might not provide enough fresh air to the building. A mechanical exhaust system ensures a proper air flow rate but increases the heating demand as the cold air enters through the envelope. A balanced ventilation system is the most efficient system, especially in combination with heat recovery, and ensures proper indoor air quality. A lot of energy can be saved by using a heat recovery system that transfers heat from the exhaust air to the supply air. However, this is only efficient if the building is airtight. If not, outside, cold air will enter the building through leaks and disturb the balanced system. The energy consumption for ventilation can be lowered by installing a heat recovery system, more efficient fans, more efficient control of the system (adjusted to use patterns instead of running constantly), a lower temperature of supply air during the heating season and by using natural cooling, such as night-flush cooling, instead of a cooling coil [5].

References

[1] Thue J V 2016 Bygningsfysikk: grunnlag. Fagbokforl

[2] Novakovic V, Hanssen S O, Thue J V, Wangensteen I and Gjerstad F O 2007 Enøk i bygninger-Effektiv energibruk *Oslo Gyldendal Undervis* 63

[3] ISO EN 2007 10456: 2007 Building materials and products–hygrothermal properties–tabulated design values and procedures for determining declared and design thermal values Brussels CEN

[4] E ISO ISO 13370:2017 Thermal performance of buildings—heat transfer via the ground—calculation methods

[5] Hens H S L 2016 *Applied Building Physics: Ambient Conditions, Building Performance and Material Properties* (New York: Wiley)

[6] E ISO 2017 15193-1 Energy performance of buildings – energy requirements for lighting – Part 1: specifications, module M9 Brussels CEN

[7] E ISO 2017 12831-1 Energy performance of buildings—method for calculation of the design heat load—Part 1: space heating load, module M3-3 Brussels CEN
[8] Standard Norge 2016 NS 3031:2016 Energy performance of buildings—calculation of energy needs and energy supply
[9] E ISO 2017 ISO 52016-1:2017 Energy performance of buildings – energy needs for heating and cooling, internal temperatures and sensible and latent heat loads – Part 1: calculation procedures

Chapter 3

Building design and envelope

Laurina C Felius, Jacob J Lamb, Fredrik Dessen and Bozena D Hrynyszyn

The first step towards energy efficient buildings is to reduce the energy demand. This can be achieved through passive and active strategies and technologies. This chapter explains how to achieve an energy efficient building envelope and which concepts are typically implemented in cold climate buildings to reduce their energy demand. Furthermore, the chapter gives an overview of passive and active components of the envelope and strategies to reduce the demand for space heating, retain heat, avoid overheating and ensure proper daylighting.

3.1 Cold climate design

One of the criteria for an energy efficient building is that it should be optimized for the building typology and local climate. By looking at vernacular and regional architecture, a lot can be learned about climatic design of buildings. With climatic design we mean the passive design (i.e. the design does not need resources to perform the intended function) of a building in response to the climate. Problems and costs can be avoided when the building design takes into account weather and climate conditions. In cold climates, the predominant style of vernacular architecture is small and compact buildings to maximize heat retention. The openings in the envelope are small, as these are weak points in the thermal envelope, so that unnecessary heat losses are avoided. Buildings were often constructed with wood, both because of the abundance in the area as well as their favourable thermal properties compared to stone. Turf was often used on roofs to improve the thermal properties. A fireplace was used as a space heating source and this was placed centrally in the building [1, 2].

As mentioned before, the focus in cold climates is to minimize heat losses through transmission and infiltration and to optimize the use of free heat gains so that the space heating demand is reduced. Five design variables that influence the energy balance of a building are the form of the building (i.e. shape, architecture, orientation and compactness), the size and orientation of the windows, the thermal

doi:10.1088/978-0-7503-3259-0ch3

properties of the building materials, the airtightness of the building and the thermal bridge value. The consequences of these parameters for cold climate design and how they can reduce the space heating demand are listed in table 3.1.

3.2 Strategies to reduce the energy demand

Strategies to reduce the energy demand can be either active or passive. A passive strategy can reduce the building's energy demand without the input of energy sources. This means that these strategies or technologies do not need energy to operate. Active strategies require energetic input to function. For this to be efficient, the energy savings need to be higher than the operational energy costs. The strategies can be divided into strategies that focus on free heating and retaining heat, strategies that focus on avoiding overheating and strategies to provide free daylight to the building. Combining these strategies results in an energy efficient building envelope. This section explains how this can be achieved in cold climates.

3.2.1 Decreasing the space heating demand

The biggest contribution to decreasing the space heating demand is by limiting the heat losses through transmission and ventilation. On a whole building level, this can be achieved by designing a compact building. The orientation of the building should be optimized for maximum solar heat gains and sheltering from prevailing winds. On a component level, heat losses can be limited by choosing high-performance insulation materials and windows, by ensuring a continuous insulation layer and by ensuring an airtight building envelope with minimized air leakage. The space heating demand can also be decreased by using active strategies such as temperature zoning. When the temperature set-point is based on the zone function and the time of day instead of a constant set-point, the zones are not heated more than needed and this decreases the space heating demand.

3.2.2 Providing free heating

Solar energy can provide free heating to the building and has a significant potential for improving the energy efficiency of buildings in cold climates. As a passive strategy, this is done by using the greenhouse effect of solar heat getting trapped behind glazed components. The system collects, stores and distributes the solar heat without energetic input. To achieve this, the system should have a collector and an element that stores the heat, often in the form of thermal mass. Utilising solar heat in the whole building can reduce the energy demand for space heating; however, during summertime it may provide too much heat and can cause overheating problems. The choice of solar gain system depends on factors such as the local climate, orientation, building function and cost. In most cases, it is best to use a combination of systems.

3.2.3 Limiting the cooling demand

Limiting the cooling demand of a building in a cold climate is not a priority for reducing the energy consumption and it is not as common to implement strategies in

Table 3.1. Effect on the energy balance of building design parameters and the consequences for cold climate design.

Parameter	Has an effect on	Cold climate design consequence
Building form, orientation and compactness	Transmission heat losses	A smaller surface area will decrease the total heat losses through the building envelope. A sheltered building will have less transmission losses because wind can enter the insulation layer and reduce the insulation properties. A proper wind barrier can prevent this.
	Heat losses through infiltration and ventilation	A more compact building (lower volume) will have less heat loss through ventilation and infiltration. Sheltered buildings will also have less infiltration heat loss as this is influenced by wind speed.
	Solar heat gains	Building form and orientation can be optimized to increase solar heat gains. Form and orientation can also be used to avoid solar heat gains so as to minimize overheating.
Window size and orientation	Transmission heat losses	A smaller window area will decrease the heat losses. However, there are requirements for minimum window area because of daylight.
	Solar heat gains	A larger window area facing south will increase the solar heat gains, though for standard windows not enough to balance out the increased heat losses. A high-performance window can have better thermal properties than opaque walls and can have more gains than losses on an annual basis.
Building materials (thermal properties)	Transmission heat losses	Improving the thermal transmittance of building materials will decrease heat losses. This can be done with better insulation materials or by increasing the insulation thickness.
Airtightness	Heat losses through infiltration and ventilation	A more airtight building has less heat loss through ventilation and infiltration. It can be achieved by installing an airtight layer. It is important that this layer is continuous around the whole building volume (no gaps at the joints).
Thermal bridges	Transmission heat losses	Fewer thermal bridges will result in fewer heat losses through transmission. It can be achieved by proper design of joints so that the insulation layer is not broken.

the building design for this purpose. Nevertheless, even buildings in cold climates may have a cooling demand during the summer. For most building types, this is solved by strategies that do not require energy, such as using natural ventilation to flush out the heat and the use of solar shading to block the incoming sun. Shading is becoming increasingly important with global temperature rising and is one of the most efficient solutions to prevent overheating. Even in cold climates, passive solar gain systems can result in overheating and an increased cooling demand during the summer. The need for shading depends on the location of the building, the size of the windows and the orientation of the windows. It is therefore essential that the shading system is adapted to the building and its local climate.

3.2.4 Providing free daylighting

Using natural light to light a building can significantly reduce the energy consumption for electrical lighting. The easiest way to achieve this is by having windows that let in the daylight. Supplying adequate levels of daylight to a building has a positive effect on the health and wellbeing of the users; however, too much daylight can result in glare and overheating of buildings. In the wintertime, large windows can become areas of heat losses and can result in discomfort by causing draft and temperature gradients in the room. In cold climates, it is therefore important that the daylight design ensures an even distribution of daylight in the room, especially in non-residential buildings. In addition, overheating due to too much direct sunlight should be prevented and heat losses through glazed areas should be minimized by either minimizing the glazed area or by using high-performance windows.

3.2.5 Creating an energy efficient building envelope

There are five main principles that need to be considered when optimizing the building envelope to be energy efficient. These may differ between climate zones, but in cold climates the main focus is to reduce heat losses and optimize heat gains while ensuring a good indoor climate.

1. Optimise the thermal envelope. For cold climates, this means minimizing the heat losses through the envelope and making optimal use of the incoming solar heat gains without creating overheating problems.
2. Ensure proper installation of the transparent building components (e.g. windows), into the opaque envelope. It is important that windows are installed as close to the insulation layer as possible and that leaks between components are avoided. To minimize heat losses during winter and overheating during summer, windows should be high-performance and with proper orientation.
3. Ensure that the construction is nearly free of thermal bridges so that local heat losses and local discomfort due to temperature differences are avoided.
4. Create an airtight envelope. This will reduce the heat losses through infiltration, and it will ensure that the thermal conductivity of the thermal insulation is as designed.

5. Install a mechanical ventilation system, preferably balanced, with heat recovery. As a result of the building becoming more airtight, it is important to have a proper ventilation system to provide enough fresh air to the building.

Combining these five principles will ensure an energy efficient envelope, though specific parameters, for example insulation thickness and window properties, should be optimized for the building type in the local climatic zone.

3.3 Components of the building envelope

3.3.1 Opaque envelope

Improving the thermal properties of the opaque envelope (i.e. walls, roof and floors) is achieved by adding thermal insulation. Due to the low thermal conductivity of insulation materials, the heat flow is significantly reduced. Commonly used insulation materials, such as mineral wool, typically have a thermal conductivity between 0.03 and 0.04 W m^{-1} K^{-1} (table 3.2). When the thermal conductivity of the chosen insulation material is known, the insulation thickness required for achieving the minimal U-value can be calculated; however, the U-value is not linear to the thickness of the insulation and the improvement of the U-value becomes less for each cm of insulation added. Figure 3.1 shows the U-value as a function of the insulation thickness for four thermal conductivity values. The figure illustrates that to achieve low U-values (below 0.1 W m^{-2} K^{-1}) a thickness of at least 300 mm (but more commonly 400 mm) is needed, unless a material with a low thermal conductivity is used.

High-performance insulation materials have a thermal conductivity that is a factor 5 to 10 lower than common insulation materials [5]. Use of these materials enables thin building envelopes with good thermal properties. This is especially a benefit in case of space constraints, height restrictions and in retrofitting projects. The thermal conductivity of a material is the sum of the conductivities for solid state, gas, radiation, conduction, convection, coupling and leakage. The conductivity for

Table 3.2. Thermal conductivities of common and high-performance insulation materials (data from [3–6]).

Insulation material	Thermal conductivity [W m^{-2} K^{-1}]
Mineral wool	0.030–0.045
Cellulose	0.040–0.055
Expanded polystyrene (EPS)	0.025–0.040
Extruded polystyrene (XPS)	0.030–0.040
Polyurethane (PUR)	0.020–0.030
Vacuum insulation panels (VIP)	0.002–0.004
Gas filled panels (GFP)	<0.040
Aerogel	0.012–0.020
Nano insulation materials (NIM)	<0.004

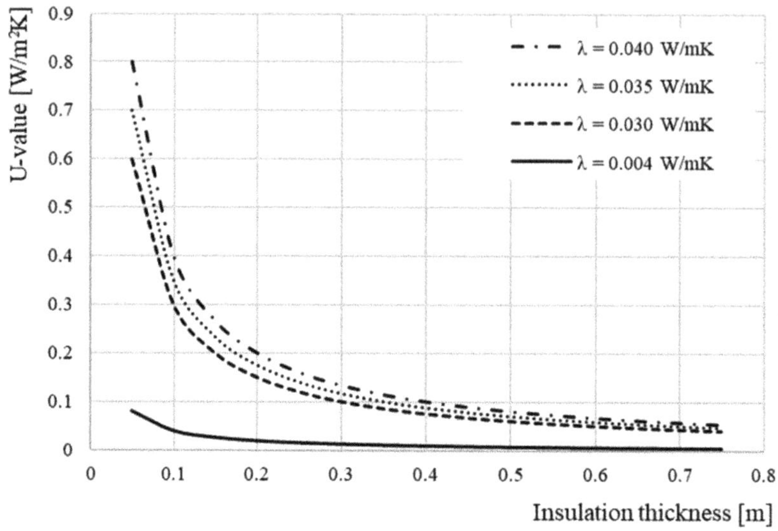

Figure 3.1. *U*-value as a function of the insulation thickness.

leakage is normally neglected as well as the coupling conductivity as it is complex. The other conductivities have to be minimized in order to reach the level of high-performance insulation materials.

Vacuum insulation panels (VIPs) and gas filled panels (GFPs) use a similar technology with an envelope and a core filled with a vacuum or gas. The thermal conductivity of VIPs is lower than that of GFPs because vacuum is a better insulator than gasses; however, maintaining a vacuum in the panel core requires more strength than maintaining a gas. Both these products have some serious drawbacks due to how they are constructed. The panels are fragile and penetration results in a significant increase of conductivity. The thermal conductivity also increases over time due to vapour and air diffusion inside. The panels cannot be adapted on-site, which can result in thermal bridges close to variations in the geometry, such as windows or corners. Because GFPs are so similar to VIPs but with worse thermal properties, they are not used. VIPs are available on the market, though it is not an optimal solution due to all the disadvantages and the high cost [5, 6].

Aerogel is another high-performance material that is available on the market, though still expensive. This insulation material is made from dried gel with a high porosity and a very low density. The main benefits are that aerogel has a relatively high compression strength, it is lightweight and can be adapted on-site. The product can also be made as a translucent or transparent material and therefore has potential to be used in windows. However, aerogel is fragile due to its low tensile strength and the product is expensive to produce.

Another product group that shows a large potential is nano-insulation materials (NIMs). NIMs are materials with a small pore structure, either open or closed, and a very low thermal conductivity. This can be achieved because of the Knudsen effect, which happens when the diameter of the material pores is smaller than the length of the mean free path of the gas molecules in the pore. This means that a gas molecule hits the pore wall more frequently than it hits another gas molecule, so the thermal conductivity for gas becomes very low. Therefore, the total thermal conductivity of NIMs significantly decreases, even if the pores are filled with air [7]. The benefit of NIMs is that the penetration of other gases and moisture does not influence the thermal conductivity as with VIPs and GFPs. The material can also be adjusted on-site, which makes it flexible in use. However, NIMs are not on the market yet though they show a lot of potential for future building design.

3.3.2 Transparent envelope

Transparent components, such as windows, play an essential role in the indoor comfort of buildings. These components form a connection between the building and its environment and influence the wellbeing of the occupants. From an architectural and health point of view, it may be desirable to have many windows, but from an energy point of view, it is more complicated. Transparent components affect the energy balance in two ways: through heat losses and through solar heat gains. In cold climates and for standard windows, the total annual heat losses are significantly higher than the solar heat gains. Glazing generally has a higher U-value than opaque elements, which means that it significantly contributes to the total heat losses of the envelope. The joints between transparent components and the rest of the envelope can also result in air leaks and thermal bridges if not installed properly [3]. The effect of the thermal properties of transparent components increases when the total energy consumption of a building decreases. To minimize the heat losses through transparent components, the thermal performance, orientation, window size and shading system should be optimized [8]. In cold climates, it is recommended to have a minimal window-to-wall ratio and a higher number of windowpanes per component [9]. Decreasing the size and thermal transmittance of the window frame also has a big impact on the heat losses of the component [10].

Although heat losses normally outweigh the solar heat gains, solar gains is an important aspect of energy efficient design through passive strategies. The total solar heat gains depend on the solar heat gain coefficient (SHGC) of the glazing: the fraction of solar radiation that is transmitted into the building. Windows with an optimized SHGC and U-value can act as net-energy-gainers in comparison to an opaque component. For example, a 4-pane window with a U-value below 0.4 W m^{-2} K^{-1} and a SHGC below 0.5 performs better in terms of the combined heating and cooling demand than a wall with a U-value of 0.15 W m^{-2} K^{-1} [8]. If the objective is to reduce the heat losses without taking into account excessive solar gains, the best results are achieved by windows with low U-values and a high SHGC. Windows with a U-value of 1.2 and 0.8 W m^{-2} K^{-1}, and a SHGC higher than 0.6 and 0.35, respectively, perform better than an opaque wall ($U = 0.15$ W m^{-2} K^{-1}) [8]. The latter values correspond with a 3-pane window. However, it should be mentioned that these windows can be too expensive. High-performance windows also increase the risk for overheating and can increase the cooling demand of a building.

Dynamic windows can adapt to the environment by changing their properties, such as the solar transmittance. The windows can either respond autonomously to changes in the environment, such as light or heat, or they can respond actively to electrical stimuli that can cause the optical characteristics of the glazing to change. The potential of these so-called smart windows in cold climates is that they can adapt their properties to reduce the cooling demand in the summer while maximizing the solar gains during the winter. The main drawbacks of dynamic windows is the high production cost and the lack of knowledge about these products [11].

3.3.3 Airtightness

Improving the airtightness of a building is essential to reduce the space heating demand, as heat is lost through cracks and leaks in the building, and to prevent thermal discomfort, such as draft. Air leakage also reduces the performance of insulation materials, resulting in a lower thermal resistance of the building envelope. An airtight building is also required for a well-functioning balanced ventilation system and a proper indoor air quality. Air leaks cause an imbalance in the ventilation system and can cause the heat recovery to be less efficient. The leaks also cause pollutants and other particles to enter the building, which worsens the indoor air quality [3, 12].

Airtightness is achieved by combining an airtight, vapour-open layer (wind barrier) on the outside and a vapour barrier on the inside of the envelope. If no vapour barrier is installed, moisture can penetrate into the construction and cause damage. It is important that these two layers are installed where they cannot be easily damaged or punctured by the user. The wind barrier should be placed so that the envelope cladding is still ventilated if necessary. Particular attention should be given to airtightness at construction joints and around openings in the envelope, as this can be complex to solve.

3.3.4 Thermal bridges

Critical thermal bridges occur around construction joints and around openings in the envelope. They can be identified using a thermography camera, which shows the temperature of the surfaces. Thermal bridges are often the result of a decreased thickness of the insulation layer or of missing insulation, but they also occur when the construction is penetrated by a material with a much higher thermal conductivity (e.g. a metal beam in a wooden construction). They can cause local discomfort due to lower surface temperatures, which can result in draft and asymmetric heat radiation, and an increased risk for moisture damage due to condensation on the interior cold surface. A thermal bridge is considered to be critical when the total heat losses increase significantly, when the difference in surface temperature causes discomfort or when the difference in surface temperature causes damage to the construction through moisture damage or fractures [12]. The risk for moisture damage increases when the air change rate decreases, for example when the airtightness is improved without having a mechanical ventilation system [3].

Figure 3.2. Passive solar systems showing a sunspace (left), Trombe wall (middle) and direct gain system (right).

Thermal bridges can be avoided by proper design of construction joints and by proper execution during construction.

3.3.5 Solar systems

Passive solar systems are categorized according to the relation between the collector and the storing element. There are three main types of passive solar systems: direct gain, Trombe wall and sunspace (figure 3.2). The simplest and most used strategy is direct gain. Every glazed area that faces the Sun functions as a direct gain system, though depending on the orientation and thermal properties they may in balance lose more heat than they gain. In order to store excess solar heat and prevent overheating, thermal mass should be used on the inside of the building. Thermal mass is related to the heat capacity of the material. When thermal mass is used in combination with a direct gain system, it is most effective and convenient to use it in the floor. This surface receives the most sunlight and a warm floor is perceived as comfortable. Another alternative is to use thermal mass in the wall facing the glazed areas or by placing elements with a high thermal mass, such as a water tank, in the room. The thermal mass effect is most efficient when a material with high thermal mass is used, such as concrete, and is not covered by materials that decrease the absorption properties [1, 2].

In direct gain systems, the collector is the whole room, which means that all objects, such as furniture, located in the room can act as thermal mass; however, only surfaces that receive direct or reflected sunlight act as thermal mass, so objects alone are not enough. This also means that it is beneficial to use light and reflective surfaces in the rest of the interior. In a direct gain system, the thermal mass surface should preferably be between 10 and 15 cm thick, but the surface area is more important than the thickness of the material.

A Trombe wall is a passive solar gain system that indirectly heats up the building. In this system, thermal mass is placed behind a glass wall of equal size to receive maximum solar radiation. If the wall is thick (30 cm), the time delay is long, and the absorbed heat does not radiate to the interior until the evening or night. In this way, the thermal mass functions as a radiator the whole night. The thermal mass can consist of water in transparent containers with a dark, non-reflective coating.

A Trombe wall is opaque and does not provide daylight to the room, unless transparent containers with water are used. Therefore, this system is often used in combination with a direct gain system that provides heat and daylight during the day.

In a sunspace, the heat gets collected in a room separate from the main building but connected through openings and vents. Sunspaces are quite popular because of their temperature on a sunny winter day; however, the temperature inside fluctuates significantly due to the large glazed area. Therefore, it should be possible to close-off the Sunspace from the rest of the building, to prevent extra heat losses during night or on cold cloudy days and to prevent overheating during warm days. The collected heat is either moved directly to the building through the openings or it is stored in the thermal mass in the sunspace. Sunspaces should be unconditioned spaces because their poor thermal properties would otherwise significantly increase the heating and cooling demand of the building. Extreme overheating can be prevented by properly ventilating the sunspace to the exterior. The thickness and area of the thermal mass depends on the purpose of the sunspace. If the sunspace heats up the house almost immediately, little thermal mass is required. If the heat is stored so that the house is heated at night, more thermal mass is needed to create the time delay [1, 2].

Thermal mass incorporated into solar heat systems can significantly contribute to lowering the energy demand by reducing the peak load as a result of overheating and by reducing the heating demand in the winter. Direct gain systems are the most flexible systems and are the most affordable; however, the thermal mass effect is not optimal as it is often blocked by objects in the room and overheating can occur as the room itself is the collector. A Trombe wall exploits the thermal mass effect to its full potential, but it does not provide daylight. Sunspaces can provide extra living space and can function as a greenhouse. However, the temperature inside sunspaces can fluctuate extremely and these systems are expensive to build. Because these types of systems are passive, they require less maintenance then active systems, which makes passive solar heat systems very durable and reliable. In cold climates, it should be taken into account that windows do not only result in heat gains, but also in heat losses. In most cases, the yearly heat losses will be bigger than the yearly heat gains. The heat losses can be minimized by using high-performance windows or by using night insulation over the glazed areas. The potential overheating problems that can arise during the summer should also be considered, even in cold climates [1, 2].

Active solar systems for reducing the energy demand, specifically for space heating, work similarly to passive solar systems. The difference is that the active systems use mechanical components, such as fans, to collect or distribute the heat. The most common active solar system that can reduce the energy demand of a building is the solar collector. This system has a collector in which a fluid is heated up by solar energy. The warmth of this fluid is transferred to the space heating system for immediate use or stored for later. This heat is mostly used to heat water, such as domestic hot water, swimming pools and commercial hot water, but it can also be used for space heating if the heating system is waterborne. A solar collector is most effective to heat up water to low operating temperatures, such as swimming pools and floor heating. It is also a good source for heating up domestic hot water because of the constant demand, though the efficiency is lower. Another efficient use for solar collectors is to preheat the supply air of the ventilation system, especially in cold climates and for buildings that require constant ventilation. These systems can be rather simple and inexpensive because the air does not have to be heated much as any increase from the outdoor temperature is beneficial. By using an active solar system as described, the energy demand for domestic hot water, space heating and heating in the ventilation system can be reduced. However, it is not cost-effective to design a solar system to cover 100% of the energy demand because solar energy is an irregular energy source. Therefore, solar systems should always be used in combination with traditional heating systems [1, 2].

3.3.6 Shading systems

Shading is used to block the incoming solar heat to prevent the building from overheating. In general, there are three type of shading: external, internal and integrated in the glazing. The most efficient shading is external, though they have the biggest impact on the aesthetics of the building. The challenges for shading systems in cold climates are to block the summer sun and let in the winter sun, while maintaining maximum view and daylight. A large horizontal overhang is more desirable than screens or louvres because of this. However, an overhang increases wind and snow loads on the building, and they need to be wider than the window to block out the Sun during the whole day. For east and west facing windows, the overhang would have to be extremely long to shade the complete window due to the low angle. Screens or louvres do not have these issues and have the additional advantage that they can be dynamic.

Shading systems can either be passive, such as fixed systems or manual operated moveable shading, or active, such as automated dynamic shading. The main advantages of fixed shading is that it is simple, low cost and does not require much maintenance; however, it cannot be adapted to the weather conditions and should be adapted depending on the orientation of the window. In cold climates, there are often large variations of the Sun angle between winter and summer and the weather conditions can be unstable. To make most use of the solar energy, it is desirable to let in the Sun during cold days but be able to block it when a risk of overheating occurs. There are two simple shading solutions that are effective in cold climates and require minimal effort. It is possible to strategically plant vegetation in front of the building that blocks the summer sun and creates minimal shadow after dropping its leaves in the fall. Another solution is to install an overhang that blocks out the summer sun, when the angle is high, but does not cast shadows when the Sun angle is low. This can also be achieved by simply removing the shading device after the summer. Dynamic shading responds better to these variations and ensures that the Sun can be blocked regardless of the time of year and weather conditions. Dynamic systems can range from very simple (adjusting twice per year) to extremely complex (continuous adjustments depending on the current situation). The latter should be automated to provide optimal control of the shading device. Automated shading systems measure (and sometimes predict) the outdoor and indoor conditions and respond accordingly. They can be integrated with HVAC systems to optimize the control of indoor comfort. Automated shading systems are more efficient than manual control but can sometimes clash with the preferences of the user and can become a disturbing factor in our environment [1, 2].

Shading systems integrated in the glazing or on the inside of the building are less efficient in keeping out the heat, though they are also less visible from the outside. They can take the form of a reflective coating, tinted glass or a shading device between the glass panes. Tinted glass or coatings are less desirable in cold climates because they block out the same amount of sunlight throughout the whole year. Dynamic windows address this issue, though this is an expensive solution. Shading devices between the panes should be automated and require an increased thickness of the window, which makes the window more complex and expensive. The space between the glazing should ideally be ventilated to remove the hot air. An example of this system design is a double façade [1, 2].

Internal shading, such as curtains, screens and blinds, is probably the most used in buildings as it is cheap, easy to install and can be easily regulated for each individual window. From an energy point of view, these are the most inefficient types of shading. Another disadvantage is that they obstruct the view; however, interior shading is not only used to block the solar heat, but also to prevent glare and for privacy. Internal shading has the potential to reflect daylight further into the room and so reduce the need for electrical lighting, and to provide insulation by adding an extra layer in front of the window [1, 2].

3.3.7 Passive cooling systems

Passive cooling strategies can be used to achieve a comfortable thermal environment during the summer. In cold climates, this can be achieved most of the time by avoiding heat gains through shading systems. Passive cooling strategies are much more dependent on the climate than heating strategies [1, 2]. In cold climates, the cooling demand can be reduced by using ventilation strategies. One strategy is to increase the airflow rate around the occupants, to increase evaporative cooling from the skin. Another strategy is night-flush ventilation. With this strategy, cool air enters the building at night to flush out the excessive heat. This is most efficient in combination with thermal mass, which stores the excessive heat during the day. Both strategies can be achieved with natural cross-ventilation (i.e. opening windows on opposite sides) if there is a sufficient pressure difference between the building sides caused by buoyancy. If this is not the case, the airflow rate can be increase through forced ventilation.

3.4 Building retrofitting

The European Union estimates that 75% of the building stock is energy inefficient, which means that it was built during a period with no or minimal energy performance requirements [13]. As buildings represent 40% of the total European energy consumption, it can be concluded that energy retrofitting of existing buildings has a huge energy saving potential. In energy renovation, the main goal is to improve the energy performance of the building. Because of this potential, the Energy Efficiency Directive (EED) states that a yearly renovation rate of at least 3% of the floor area of government owned and occupied buildings should be achieved [14]. In addition, all buildings that undergo major retrofitting should fulfil the national energy performance requirements of the Member State [15]. Major renovation is described as when 25% of the building envelope or 25% of the value of the building is renovated. This requirement also counts for all buildings over 1000 m^2. However, retrofitting government owned and occupied buildings alone is not enough. About two-thirds of the share of the building sector's energy consumption comes from residential buildings [16].

In addition to energy savings and a reduction of GHG emissions, there are other global and national benefits from retrofitting. Energy prices can be reduced and less fluctuating, because the energy demand is lower and more stable. This will also result in a better energy security as the load on the energy supply systems decreases and the pressure of producing energy from natural resources is reduced. A lower energy consumption also means that less fuel has to be imported, the export of fuel can be increased, and expenses related to government buildings are reduced. Retrofitting on a large scale creates a new job and has macroeconomic benefits, such as an increased GDP and increased spending on other products [17]. Though benefits on a global and national scale are important on a large scale, they may not be convincing for building owners and users. However, there are plenty of other benefits from building renovation that appeal to the individual. These benefits are listed in table 3.3, divided into benefits from a building owner perspective and from a user perspective. In privately owned buildings, the building owner and user can be the same person (e.g. housing).

The building owner can decide to combine energy retrofitting with regular maintenance and repairs. Instead of bringing the building back to its original

Table 3.3. Retrofitting benefits from a building owner and user perspective [17].

Building owner perspective	User perspective
Reduces the operational costs for the building	Reduces the energy costs for the user (energy bill)
Reduces the time and workload for building operation and maintenance	Improves the level of comfort in the building, including better air quality, temperature control and acoustics and fewer drafts and less mould growth risk
Improves the productivity of the building occupants (non-residential context)	Improves overall health and wellbeing of the user
Increases the value of the building	Increases the disposable income as a result of reduced energy costs

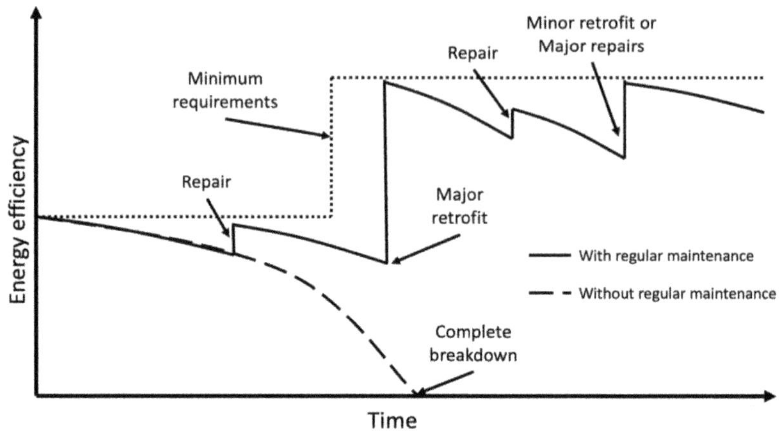

Figure 3.3. Energy efficiency of a building over time.

energy performance, the building is upgraded to reach a more ambitious level. This may also be necessary when the minimum requirements in the building code have become stricter, as is illustrated in figure 3.3. When maintenance and retrofitting are combined, the costs for retrofitting are reduced by the costs otherwise needed for maintenance. Retrofitting often requires that the cladding of the building envelope is temporarily removed. If the cladding needs to be replaced and the envelope is extra insulated at the same time, there are no additional costs for removing and replacing the cladding allocated to retrofitting; however, if insulation is added while the cladding has not reached the end of its service life, the existing cladding has to be removed and placed back (or replaced too early), resulting in extra costs.

3.4.1 Energy savings and cost-effectiveness

Energy retrofitting in cold climates can result in significant energy savings. The highest energy savings are achieved when a holistic approach is used, as illustrated in figure 3.1 [3, 18]. Felius *et al* (2019) gave an overview of residential case studies in cold climates that evaluate the potential of energy retrofitting [3]. Other studies presented retrofitting case studies in various climates for residential buildings [19], historical buildings [20] and various building typologies [21]. Achieved savings are highly dependent on the local climate, the state of the building, building constraints and funding. With an unlimited budget and no space constraints, the building can be retrofitted to a zero-energy building; however, this is often not the case.

The total investment costs can be reduced when retrofitting measures are combined, such as upgrading the walls at the same time as the windows. Another way to minimize the investment costs is by combining retrofitting measures with scheduled maintenance, as mentioned earlier. The cost-effectiveness is evaluated by comparing the investment costs with the outcome in the form of energy savings, including the cost of energy, maintenance and operation. It is influenced by the energy price, interest rate, energy savings, lifetime of the components and investment costs. Simple methods can be used to quickly estimate the cost-effectiveness, such as

Table 3.4. Overview of literature on cost-effective retrofitting case studies with the building typology and climate zone.

	Building typology				Climate zone		
	Single-family housing	Multi-family housing	Offices	Educational buildings	Other	Cold climate	Various
Alev *et al* (2015) [22]	x					x	
Dodoo *et al* (2017) [23]		x				x	
Holopainen (2017) [24]		x				x	
Kalamees *et al* (2017) [25]		x				x	
Kurnitski *et al* (2014) [26]	x	x	x	x	x	x	
Liu *et al* (2014) [27]		x				x	
Mata and Johnsson (2017) [28]	x	x	x	x	x	x	
Niemelä *et al* (2016) [29]				x		x	
Niemelä (*et al*) 2017 [30]		x				x	
Pikas *et al* (2014) [31]			x			x	
Zangheri *et al* (2018) [32]	x	x	x	x		x	x

the payback period or net present value. For a more accurate evaluation of the cost-effectiveness, the life-cycle cost method is commonly used. Table 3.4 lists several case studies that focus on energy and cost-effective retrofitting in cold climate.

Significant energy savings can cause a rebound effect. This means that the reduced energy demand caused by energy retrofitting is cancelled out by an increase of energy use because the saved money is invested in services and products that use energy. Instead of accepting energy efficiency as receiving the same product for less energy, the user chooses to receive more product for the same amount of energy as before the retrofit (which is still energy efficient, though not beneficial for lowering GHG emissions). For example, when the energy cost is cheaper than before the user may choose to turn up the heat, which increases the energy demand for space heating. Another example is that because of lower energy costs, the production of goods can be increased, resulting in increased energy consumption. More available income can also lead to an increased energy consumption of other producers, such as when a family uses the savings from a lower energy bill to go on an extra vacation (by plane). Although the rebound effect exists and can influence the estimated energy savings, it is not strong enough to cancel out the energy (and financial) savings that are achieved through energy retrofits [17].

3.4.2 Challenges

Though the benefits of retrofitting seem clear and convincing, there are still many challenges in retrofitting projects. The main global challenge when it comes to retrofitting is increasing the annual retrofitting rate. This rate is currently estimated at 0.6% for residential buildings in the European Union and is expected to increase to 1% by 2030 [33]. This needs to increase further to reach the European targets. Other challenges include those for starting the process and challenges during the renovation and can be divided into challenges on a national or individual scale and practical challenges during the retrofitting. The most essential challenges for building retrofit are summarized in table 3.5.

The Energy Efficiency Financial Institution Group (EEFIG) has proposed five strategies to address challenges related to retrofitting. All retrofitting benefits should be made clear to key decision makers, such as public authorities and building owners. Energy performance certificates, building codes and the enforcement of those should be improved so that it is easier to identify, measure, report and validate the energy savings potential of buildings. It should be easier for decision makers to access the correct and relevant data needed to understand the benefits from retrofitting and choose a suitable solution. There should be standards for all elements of the process, including contracts, procedures, verification, reporting and others. Last but not least, there should be proper use of European funding schemes to increase the available investment amount and to help engage the finances of the private sector [16].

3.4.3 Common retrofit solutions

Many of the mentioned passive and active strategies are easy to implement during the design process; however, when a building undergoes energy renovation the situation is more complex. In cold climates, energy retrofitting should focus on the same principles as described in the introduction (see figure 1.1). As the space heating demand dominates the total energy consumption, the first step is to minimize heat losses through the envelope. After this, the energy consumption should be reduced by installing energy efficient HVAC systems with control systems. These principles are the same as for new buildings and the only difference is that the building is no longer in a design phase but is already built. This often causes challenges for implementing energy saving measures. This section only mentions common retrofitting measures and challenges to reduce the space heating demand of a building. This is a summary of a more extensive discussion about retrofitting measures for residential buildings in cold climates (given by [3]). Practical solutions have also been presented for adding insulation to different types of constructions [12].

One of the most effective retrofitting measures to decrease the space heating demand is to improve the thermal properties of the building envelope [3]. This can be done by adding insulation to the opaque components and by switching to high-performance windows. This not only reduces costs by reducing the energy consumption, but also because HVAC systems can be downsized as a result of a lower space heating demand. When adding additional insulation, three problems can arise:

Table 3.5. Challenges in different stages of the retrofitting project and on different levels [16].

National scale (e.g. government)	Individual scale (e.g. building owner or user)	Practical (e.g. during planning and construction)
Building codes and standards Building codes and standards should support investment in energy efficiency measures. Lack of adjusted energy performance requirements for retrofitting projects. Enforcement of the codes and standards with penalties for failing to fulfil requirements. Lacking standards for measuring, reporting and verifying the achieved energy efficiency.	*Costs* Difficult to justify large expenses without taking into account other benefits, such as a longer building lifetime and increased comfort and health. No fiscal benefits for investing in energy efficiency measures. Loans for building improvements are person-bound instead of building-bound. Lack of financial resources to pay for technical expertise.	*Indoor comfort* More airtight building means natural ventilation is not sufficient. Moisture damage can occur if not retrofitted properly. Leaks and thermal bridges can occur if the building is partly renovated. Existing HVAC systems can become under- or over-dimensioned.
Funding The EU has available funds for retrofitting, but it is essential that these are used in a smart way. Economic benefits from retrofitting (lower energy bill and increased property value) should be considered when giving funding/loans.	*Knowledge* Lacking awareness of decision makers (e.g. leadership, house owner) about energy cost and energy efficiency. Lack of technical knowledge and expertise when developing large projects (municipal scale) or when making decisions (house owner). Lacking understanding of risks for investors.	*Space constraints* There may not be space for ventilation ducts (low ceiling height) or technical rooms. There may not be space for extra fire safety measurements, such as escape routes.
Accessibility of information Useful data and technical expertise on key aspects in the retrofitting process is not always made accessible to policy makers and decision makers. Complex data and procedures make it more difficult to execute retrofit measures.		*Other* It can be unclear what the state of the building is (inside the walls, structural, etc). It is difficult to renovate the building while it is in use.

moisture problems, thermal bridges and space constraints. Moisture problems and thermal bridges can be avoided when insulation is added on the exterior side of the construction, though some thermal bridges are difficult to remove (such as around protruding balconies); however, adding exterior insulation is not always possible due to space constraints or if the building has a heritage status. In those cases, insulation can be added on the interior side of the construction. This can result in thermal bridges, for example where an inner wall meets the building envelope. Adding insulation on the interior side of the construction can lead to condensation in the construction. Condensation can also happen when exterior insulation is added, but this is easily avoided by adding a vapour barrier on the inside. Due to the condensation risk with interior insulation, the total thickness that can be added is limited. Insulation can in some cases also be installed in cavities, such as in between the beams of a timber framed ceiling construction or in the cavity of a brick wall. This is only suitable if the cavity is large enough and if ventilation of the cladding is not obstructed. A solution to maximize the available space is by using high-performance insulation materials. Due to their low thermal conductivities, the same U-value can be achieved with less material, which is beneficial when there are space constraints. Though these products are highly efficient, they are still expensive to produce and have other issues that should be addressed, such as the adaptability on-site. Improving the thermal properties of transparent components, such as windows, is also essential for reducing the space heating demand. In old buildings, windows result in increased heat losses through transmission and infiltration, because windows are often leaky. In retrofitting, it is therefore essential that leaks and thermal bridges are minimized when installing new windows. This can be achieved by placing them as close to the insulation layer as possible and by using sufficient sealing around the joints [34].

In addition to improving the thermal insulation of the envelope, the air leaks and thermal bridges should be removed. The airtight layer of a building, if there is one, often has breaks around construction joints. This can in some cases account for up to 30% of the space heating demand [35]. In the most ideal case, the existing airtight layer is replaced by a continuous vapour barrier on the inside of the construction; however, in retrofitting projects this can be costly and time-consuming to execute. Nevertheless, a building without an airtight layer and vapour barrier will have high infiltration heat losses and condensation problems if the insulation thickness increases. Thermal bridges can account for up to 28% of the transmission heat losses [36]. They also increase the risk for mould growth due to lower local surface temperatures. Therefore, it is recommended to remove thermal bridges by insulating from the exterior side during renovation.

As old buildings often do not have a mechanical ventilation system, fresh air is provided by the windows and through cracks. This means that when windows are upgraded and the airtightness improves, the natural ventilation rate is not high enough [34]. Therefore, it is important to install a mechanical ventilation system after retrofitting the envelope. Balanced ventilation with heat recovery is the preferred system in cold climates and will significantly lower the space heating demand of a building [37]; however, due to space constraints or low ceiling heights it may not be possible to install a balanced ventilation system. Then, ventilation systems that ensure a proper airflow rate can be suitable solutions [3].

Table 3.6. Overview of strategies and components effective in cold climates to reduce the energy demand of buildings.

Measure	Type of measure	Impact on the energy demand
Thermal insulation	Passive	Reduces the space heating demand by minimizing heat losses through transmission, including increased heat losses around thermal bridges
High performance window	Passive or active	Reduces the space heating demand by minimizing heat losses and by providing free solar heat gains. May increase the cooling demand by providing too much solar gains. Dynamic windows can change their properties to regulate incoming solar gains.
Airtight envelope	Passive	Reduces the space heating demand by minimizing heat losses through infiltration and increased transmission heat losses around thermal bridges.
Solar gain system	Passive	Reduces the space heating demand by providing free heating (internal gain).
Solar collector	Active	Reduces the space heating demand, including ventilation heat, and for (domestic) hot water by providing free heating (internal gain).
Shading	Passive and active	Reduces the cooling demand by blocking excessive solar energy (internal gain). It can be either passive (fixed or manually operated) or active (motorized control).
Night-flush ventilation	Passive or active	Reduces the cooling demand by flushing out the excess heat during the night. It can be either passive (natural ventilation) or active (forced ventilation).
Energy retrofitting	Passive and active	Significantly reduces the energy consumption of existing buildings. Energy retrofitting is a combination of the above-mentioned measures, both active and passive, with a focus on improving the thermal properties of the building envelope.

3.5 Conclusion

This chapter gave an overview of passive and active strategies and measures that can reduce the energy demand of a building, mainly focusing on the space heating demand. Table 3.6 summarizes the measures discussed in the chapter and how they influence the energy demand. The most effective solutions are to decrease transmission and infiltration heat losses through the building envelope. This can be achieved with well-insulated opaque elements and high-performance windows in combination with providing proper airtightness and reducing thermal bridges in the construction. It is important to take into account that mechanical ventilation is required to provide enough fresh air to the building and that one should be careful with solar heat gains to avoid overheating.

References

[1] Lechner N 2014 *Heating, Cooling, Lighting: Sustainable Besign Methods For Architects* (Hoboken, NJ: Wiley)
[2] Lechner H M 2009 *Heating, Cooling and Lighting* (Hoboken, NJ: Wiley)
[3] Felius L, Dessen F and Hrynyszyn B 2019 Retrofitting towards energy efficient homes in European cold climates—a review *Energy Effic.* Accepted **13** 101–25
[4] Aditya L *et al* 2017 A review on insulation materials for energy conservation in buildings *Renew. Sustain. Energy Rev.* **73** 1352–65
[5] Jelle B P, Gao T, Sandberg L I C, Tilset B G, Grandcolas M and Gustavsen A 2014 *Int. Conf. On Advances in Civil, Structural and Mechanical Engineering - ACSME 2014* 97–104
[6] Jelle B P 2011 Traditional, state-of-the-art and future thermal building insulation materials and solutions–Properties, requirements and possibilities *Energy Build.* **43** 2549–63
[7] Jelle B P, Gustavsen A and Baetens R 2010 The path to the high performance thermal building insulation materials and solutions of tomorrow *J. Build. Phys.* **34** 99–123
[8] Grynning S, Gustavsen A, Time B and Jelle B P 2013 Windows in the buildings of tomorrow: energy losers or energy gainers? *Energy Build.* **61** 185–92
[9] Thalfeldt M, Pikas E, Kurnitski J and Voll H 2013 Facade design principles for nearly zero energy buildings in a cold climate *Energy Build.* **67** 309–21
[10] Kalbe K and Kalamees T 2019 Influence of window details on the energy performance of an nZEB *J. Sustain. Archit. Civ. Eng.* **24** 61–70
[11] Casini M 2018 Active dynamic windows for buildings: a review *Renew. Energy.* **119** 923–34
[12] Novakovic V, Hanssen S O, Thue J V, Wangensteen I and Gjerstad F O 2007 Enøk i bygninger-Effektiv energibruk *Oslo Gyldendal Undervis.* 63
[13] European Commission Clean energy for all Europeans [Internet] 2019. Available from: file:///Users/jacoblamb/Desktop/MJ0319092ENN.en.pdf
[14] Parliament E 2012 Directive 2012/27/EU of the European Parliament and of the Council of 25 october 2012 on energy efficiency, amending directives 2009/125/EC and 2010/30/EU and repealing directives 2004/8/EC and 2006/32/EC *Off. J. Eur. Union L.* **315**
[15] Recast E 2010 Directive 2010/31/EU of the European Parliament and of the Council of 19 may 2010 on the energy performance of buildings (recast) *Off. J. Eur. Union.* **18** 2010
[16] Efficiency E 2015 The first fuel for EU economy Final Rep
[17] Ryan L and Campbell N 2012 *Spreading the net: the multiple benefits of energy efficiency improvements* International Energy Agency
[18] Harvey L D D 2009 Reducing energy use in the buildings sector: measures, costs, and examples *Energy Effic.* **2** 139–63
[19] De Boeck L, Verbeke S, Audenaert A and De Mesmaeker L 2015 Improving the energy performance of residential buildings: a literature review *Renew. Sustain. Energy Rev.* **52** 960–75
[20] Martínez-Molina A, Tort-Ausina I, Cho S and Vivancos J-L 2016 Energy efficiency and thermal comfort in historic buildings: a review *Renew. Sustain. Energy Rev.* **61** 70–85
[21] Ma Z, Cooper P, Daly D and Ledo L 2012 Existing building retrofits: methodology and state-of-the-art *Energy Build.* **55** 889–902
[22] Alev Ü, Allikmaa A and Kalamees T 2015 Potential for finance and energy savings of detached houses in Estonia *Energy Procedia.* **78** 907–12
[23] Dodoo A, Gustavsson L and Tettey U Y A 2017 Final energy savings and cost-effectiveness of deep energy renovation of a multi-storey residential building *Energy* **135** 563–76

[24] Holopainen R 2017 Cost-efficient solutions for finnish buildings *Cost-Effective Energy Efficient Building Retrofitting* (Amsterdam: Elsevier), 363–83
[25] Kalamees T, Kuusk K, Arumägi E and Alev Ü 2017 Cost-effective energy and indoor climate renovation of Estonian residential buildings *Cost-Effective Energy Efficient Building Retrofitting* (Amsterdam: Elsevier), 405–54
[26] Kurnitski J, Kuusk K, Tark T, Uutar A, Kalamees T and Pikas E 2014 Energy and investment intensity of integrated renovation and 2030 cost optimal savings *Energy Build.* **75** 51–9
[27] Liu L, Moshfegh B, Akander J and Cehlin M 2014 Comprehensive investigation on energy retrofits in eleven multi-family buildings in Sweden *Energy Build.* **84** 704–15
[28] Mata É and Johnsson F 2017 Cost-effectiveness of retrofitting Swedish buildings *Cost-effective Energy Efficient Building Retrofitting* (Amsterdam: Elsevier), 343–62
[29] Niemelä T, Kosonen R and Jokisalo J 2016 Cost-optimal energy performance renovation measures of educational buildings in cold climate *Appl. Energy* **183** 1005–20
[30] Niemelä T, Kosonen R and Jokisalo J 2017 Cost-effectiveness of energy performance renovation measures in Finnish brick apartment buildings *Energy Build.* **137** 60–75
[31] Pikas E, Thalfeldt M and Kurnitski J 2014 Cost optimal and nearly zero energy building solutions for office buildings *Energy Build.* **74** 30–42
[32] Zangheri P, Armani R, Pietrobon M and Pagliano L 2018 Identification of cost-optimal and NZEB refurbishment levels for representative climates and building typologies across Europe *Energy Effic.* **11** 337–69
[33] von Manteuffel B, Petersdorff C, Bettgenhauser K and Boermans T 2016 EU pathways to a decarbonised building sector
[34] Bestain Z 2016 Step by step retrofits with passive house components [Internet] [cited 2019 Oct 8]. Available from: https://europhit.eu/sites/europhit.eu/files/EuroPHit_Handbook_final_Optimized.pdf
[35] Jokisalo J, Kurnitski J, Korpi M, Kalamees T and Vinha J 2009 Building leakage, infiltration, and energy performance analyses for Finnish detached houses *Build. Environ.* **44** 377–87
[36] Berggren B and Wall M 2013 Calculation of thermal bridges in (Nordic) building envelopes–risk of performance failure due to inconsistent use of methodology *Energy Build.* **65** 331–9
[37] Dodoo A, Gustavsson L and Sathre R 2011 Primary energy implications of ventilation heat recovery in residential buildings *Energy Build.* **43** 1566–72

IOP Publishing

Energy-Smart Buildings
Design, construction and monitoring of buildings for improved energy efficiency
Jacob J Lamb and Bruno G Pollet

Chapter 4

Smart components and systems

Laurina C Felius, Jacob J Lamb, Bozena D Hrynyszyn and Fredrik Dessen

After reducing the energy demand of the building through optimizing the building envelope and reducing the energy consumption by optimizing the HVAC systems, the energy consumption of the building can be further reduced by implementing smart components and a building control system (BCS). This chapter focuses on solutions and technologies that can monitor and control the energy use of the building. The chapter also discusses the challenges and benefits from a BCS and discusses the interaction between the system and the user.

4.1 Introduction

It is economically and socially advantageous to reduce the cost of maintaining the internal environment of a building by measurement and control of various parameters, allowing the optimisation of building environments to reduce energy wastage and improve comfort. This requires sensors to provide live data monitoring from the building's environment that can be used to trigger interventions to maintain the desired environment. Significant advancements in the fabrication of semiconductors and smart algorithms have made it possible to design and develop smart sensors [1]. Furthermore, the use of wireless sensor networks (WSN) for gathering data, and logical methods for contextual interpretation and assessment allows smart sensor networks to also monitor the wellbeing of the people within the building environment [2].

After reducing the energy demand of the building through optimizing the building envelope and reducing the energy consumption by optimizing the HVAC systems, the energy consumption of the building can be further reduced by implementing a BCS. This chapter focuses on solutions and technologies that can monitor and control the energy use of the building, discussing the challenges and benefits from a BCS and the interaction between the system and the user. The overall description of a smart sensor network for buildings will also be described.

doi:10.1088/978-0-7503-3259-0ch4

4.2 Smart system description

The concept of smart buildings has been developing for several decades now and is defined as a system that can automate and assist the occupants of the building with control, automation and optimisation of the environment. Therefore, the objective of smart buildings is to increase the comfort and to make the day-to-day life of the occupants easier. This may be achieved by identifying the relevant human activity and increasing the automation, or by using remote control to provide high levels of comfort, security, energy management and reduce emissions. Overall, smart homes generally have the following fundamental characteristics:

- automation;
- multifunctionality;
- adaptability;
- interactivity; and,
- efficiency.

4.2.1 Smart building network

A smart building network consists of a variety of sensors, a network for data transfer and a central station where all data is collected and monitored. The sensors are placed within the building and can be connected to a hub (through wireless technology). This hub connects to a server computer (or central station) that can be accessed by the administrators of the network. Some sensors that can be used in a building are:

- temperature sensors;
- irradiation sensors;
- humidity sensors;
- dry contact sensors;
- magnet-based open/close sensors; and,
- passive infrared (PIR) sensors.

The configuration of sensor networks can usually be modified to suit their application (e.g. frequency of measurement).

4.2.2 Information and communication technologies

Many of the advantages of a smart building arise from the capability to provide reliable performance and responsiveness of information. The rapid advancements in communication and information handling technologies has allowed the smart building system to be fully realised. In turn, this has resulted in improved efficiency, increased automation, reliable forecasting (by monitoring behaviours over time) and safer operation of appliances within the building. This can be further extended to smart electricity grids, where usage of electricity from the grid can be optimized to reduce financial cost and emissions.

4.2.3 User behaviour

User behaviour is an important factor when considering the objective of improving the performance and comfort of a building. An example of an improvement that applies to everyone was suggested by the intergovernmental panel on climate change (IPCC), where up to 70% of energy consumption from lighting could be saved by altering behaviour [3]. Many individuals in society are becoming more conscious of their impact on the environment in recent years and are becoming increasingly interested in smart building devices and tools as a consequence. Moreover, incentives from governments and industries is beginning to drive the demand for residential energy management, an area that will significantly grow in years to come [4]. This will allow the management of energy for buildings in a way that is tailored for the user's energy usage patterns, allowing improvements to be made in energy usage to reduce the financial cost or emissions.

4.3 Smart building technology classification

Although smart building networks and technologies have been established for many years now, the directionality and complexity of the communication devices represents a real challenge to increasing the exchange of information. Home automated networks (HAN) are becoming increasingly more interesting as they have been developed to incorporate a bidirectional communication system allowing monitoring and control of a building. However, if this is to be fully realised, combinations of useful technologies with well-designed software programs are required to successfully inform, engage and motivate users via the following determining factors [5]:

- data collection;
- data processing;
- data representation; and,
- control and interaction capabilities.

Data collection technology allows the collection of all data and allows access to this data. The processing of data analyses what data is relevant and combines the various sources. The relevant data can then be represented and made accessible to the users or other technologies. Finally, the user can access the status and monitor the functions of related technologies allowing a bidirectional communication and interaction. These factors need to be considered when designing smart systems so that the data can be tailored to the end users. Collected data can be shown to users as either direct or indirect feedbacks. Direct feedbacks are generally the representation of collected data in real-time; whereas, indirect feedbacks are generally derived from post-processing of the data and provided later in a summary (e.g. daily or weekly feedback).

Data from smart network systems can be presented in a variety of ways (e.g. energy usage, peak consumption, predicted cost and environmental footprint). This can then be compared to benchmark values and trends to give feedback on the user behaviour over time. To be effective, feedback that is given to the end-user must be:

- direct;
- personalised;
- comparable; and,
- flexible.

By being direct, the information is rapidly disseminated to the user, making it more effective for use by the user; however, this may require a certain degree of knowledge of the variable from the user. This can be developed with personalisation, where the end user receives data that is useful to them specifically. This can allow comparison of their data with their own historical data, or with others. Finally, the technology needs to be flexible so that it can be continually improved to meet the developing requirements of the end user.

The two main approaches used to communicate the information to the user can be characterised as systematisation and gamification. Systematisation is the characterisation of a sector or sensor into its own system, and data retrieved from this sector or sensor is presented at the level of that individual system and compared with the average and historical data from that system or other similar systems. Gamification is the way of presenting the data using elements and concepts typically used in computer games. This increases the intuitive interaction of the user with the technology, generally through a graphical user interface (GUI).

4.4 Smart building technologies

With regard to the technologies for smart buildings, there are four main groups that the technologies can be grouped into: integrated wireless technologies (IWT), home energy management systems (HEMS), smart home micro-computers (SHMC) and home automation (HA).

4.4.1 Integrated wireless technologies

Several investigations on information and communications technologies (ICTs) for smart grids and smart buildings can be discovered in the literature [6]. IWT is a frequently used communication within an office building, a personal home or any other residence to enable short-range inner and external communication across smart home technology. Wired techniques are often not preferred, with IWTs being the preferred option. For many smart grid applications, the use of wired solutions would be economically and/or physically prohibitive. Wireless technologies instead provide advantages such as reduced hardware and assembly costs, rapid deployment, extensive availability and higher flexibility. Furthermore, IWT systems for remote monitoring and control of home appliances could be implemented via a GUI. It enables inclusion and communication within house energy management technologies, whereas when incorporated into a smart grid, IWT has some disadvantages. Local renewable energy generation devices like photovoltaic (PV) modules are not presently included. The main wireless communication technologies and network protocols currently deemed suitable for smart building use are outline bellow.

Bluetooth is a wireless communications system that can exchange data over short distances (up to 15 m). This is achieved by employing short-wavelength radio in the range of 2400–2480 MHz. It has a low energy requirement, fast transfer of data (up to 25 Mbps) and a widespread availability.

DASH7 is an open source wireless network protocol for sensors and actuators. It operates in the 433, 868 and 915 MHz radio bands with ranges up to 2 km without requiring large amounts of power. It has 128-bit AES encryption with data transfer speeds of up to 200 kbps and can penetrate concrete and water.

RFID (radio frequency identification) is a system where the ID of the objects is transmitted wirelessly by radio waves. It is characterised on the frequency range used (low: 124–135 kHz, high: 13.56 MHz and ultra-high: 860–960 MHz). the range of transmission is anywhere from 30 cm (for low frequency use), to 15 m (for ultra-high frequency use).

Wi-Fi is an extremely popular technology used in many electronic devices. The main advantage of Wi-Fi is the extensive support that is already existing, and large utilisation network. The added advantage is that most buildings will already have Wi-Fi networks in place, reducing the initial cost of setting up smart systems based on Wi-Fi. It can also communicate with the internet itself as it has internet protocols built into the technology. The transfer of data over Wi-Fi can reach up to 300 Mbps, with a range up to 100 m.

ZigBee is a wireless mesh network that is very economical and efficient. It can send low amounts of data to a personal area network (40–250 kbps). Its range can be from 10–75 m. Its uses include device control, messaging, automation, electronics and remote monitoring.

4.4.2 Home energy management

Because of the energy shortage and the impacts of global warming, development of the household energy management system (HEMS) started. It has become a common research topic since its first implementation in 1976 [7]. The HEMS enables automatic management and control of a building's energy use and helps decrease peak demand for electricity charges for customers and users [8].

In regions of North America and Europe with a high latitude, the amount of HEMS facilities is increasing due to the number of hours of darkness each year. In these regions, the HEMS considerably reduces the complete requirement for energy. The peak energy load can be decreased by an estimate of 30% and the operating cost of electricity by 23% [9, 10]. The use of energy management systems in intelligent grids has become a concern to distribute locally the energy produced by customers according to the energy prices, which are governed by regular rates. Future trends could be based on specific hourly prices (i.e. inclining block rates (IBR), real-time pricing (RTP), time of use (ToU) and critical peak pricing (CPP)).

If this technique is used, the HEMS will enhance the number of products that customers could use to conduct regular duties. These include such products as watching power utilization data, regulating thermostats and viewing their account and billing data. All forms of products in the housing sector can be split into three classifications (control equipment, graphical user interfaces and enabling technologies). These are covered in table 4.1.

HEMS are advantageous because they increase the savings for users and utilities providers, reduced peak energy loads, include local renewable energy sources, allow the building to be inserted into the grid as a separate local grid and they allow for historical comparisons of building utilisation of energy. It is recommended to use a unified control function for the user interface, which involves a command for the use of devices, and it would be helpful to inform end-users on how to use integrated automated control systems to prevent automation whose correct function end-users do not understand. While it would be essential to create a tailor-made user interface for a HEMS to function in intelligent households, HEMS would allow the use of locally produced electricity, power conservation inclusion and an effective link to a smart grid. Table 4.2 presents a comparative of the analysed HEMS accessible on the market.

4.4.3 Smart building micro-computers

Smart home micro-computers (SHMC) are small pcs linked to other appliances to automate and regulate the entire smart building scheme. They are composed of a supplementary component and a microcontroller that facilitate programming and integration into other circuits. Their standard connectors are a significant element, allowing consumers to connect to a central processing unit (CPU) panel with a variety of interchangeable add-on modules referred to as shields. They enable people

Table 4.1. Home energy management systems.

Category	Technology	Description
Control devices	Centralised	Allows multiple appliance communication with users able to manage from a single location.
	Device/appliance level	Control of a single device or function by the user.
	On-board appliance automation	Control the functionality of a device.
GUIs	Building energy display	Stand-alone display in the building
	Web portal	Online interface accessible from anywhere via the internet.
	Phone application	Device specific interfaces for smart phones.
Enabling	technologies	Sensing
	Dynamic variable acquisition within the building.	
	Communications	Network support devices.
	Communication protocols	Standards that allow the communication between individual nodes on a network.

Table 4.2. Home energy management systems comparison.

Product	Main features and strengths
CISCO	Energy management for all sizes of buildings. Has advanced programming to maximise energy use reduction and cost savings.
DigitalSTORM	Connects all electronic devices through existing power lines within the building and includes a smart meter and a control system. There are many applications that use this system.
Dreamwatts®	A web-enabled and user-friendly HEMS for monitoring and control of buildings. Is open source and accessible from anywhere with an internet connection.
e-GOTHAM	An open smart-grid system that gives the user the ability to manage components within sections of the building grid to operate in collaboration with the smart grid.
EmonCMS	Is a free application for logging visualising and processing energy usage, temperatures and environmental data.
Energy Team's Energy Data Collector Tool	Consists of an integrated web server for configuration and data analysis, and a gateway to collect, store and send data to the central server. The energy usage can be then visualised through a web-based application.
Google PowerMeter	An open source energy monitoring utility with energy usage visualisations, information sharing and tips for saving energy.
Savant	Allows controlling and setting of energy usage from a GUI on an electronic device.
SMARTHEMS™	An artificial intelligence-based system that uses a smart energy gateway to manage energy storage, generation and savings.
The Energy Navigator platform	A web-based system for building management. Automatic reporting informs the user about performance as well as issues within the system.

to create interactive projects and apps with the environment by using various extensible connectors and getting feedback from various detectors and by regulating lamps or other actuators affecting their surroundings. There are some instances of apps in literature where SHMCs were coupled with wireless detectors to generate home automation devices for monitoring and controlling home appliances [11]. Some SHMCs include Arduino, Banana Pi, Raspberry Pi, BeagieBone Black and Libelium Waspmote. Generally, they are all able to use open source programming and have large online support communities for a huge variety of applications. The drawbacks of these systems are that they generally require the user to have some technical background, basic electronics knowledge and time required to set them up.

4.4.4 Home automation systems

Home automation (HA) offers a smart tool for monitoring and learning practices of users and for anticipating and facilitating their motions. By communicating with users directly, HA can simplify the systems and make them more convenient or provide some energy efficiency savings [12]. HA is a component of the smart home management scheme. However, for user interaction, the HA scheme should be coupled with non-automated systems. Using only HA schemes, for instance, would not allow customers to change their power use. If so, feedback can be given to end-users as part of the smart home automation scheme depending on the control operations conducted. Such technology could well be incorporated into a smart scheme that reduces energy and increases building warmth and visual comfort by introducing both short-term and long-term heat and visual indicators of discomfort [13, 14].

4.5 Intelligent buildings

A rather recent development in the building sector is the transition to smart or intelligent buildings. The concept itself was first introduced in the end of the 20th century. There is no fixed definition for intelligent buildings and existing definitions vary from *buildings that respond to user requirements* [15–17], to *buildings that 'integrate and optimize building structures, systems, services and management to create a productive, cost-effective and environmentally approved environment for building occupants*' [18, 19]. This definition may be extended by including the term 'ability to learn from the user' [20]. Intelligent buildings are not only able to react and change according to user behaviour, but they should also be able to learn and adapt in order to constantly strive for optimal performance. The authors of this book believe that a smart building is an energy efficient building that is able to interact with the user.

Intelligent buildings are controlled by a BCS, which originally was designed to minimize the energy use for building operation. In addition to efficient use of the operational energy, a BCS can also achieve energy and cost savings, a high level of indoor comfort and a high level of indoor air quality [21]. To accomplish these results, the BCS needs to be able to control HVAC systems, shading systems, lighting systems and window opening. A BCS can also control elements that do not influence the energy performance, such as building security. A BCS performs three essential tasks to achieve optimal control of the building. It minimizes the effect disturbances have on the desired output variable, it minimizes the difference between the desired output variable and the set-point variable, and it minimizes the reaction time between the detection of an error (deviation) and a system adjustment [22].

4.5.1 Challenges and opportunities

Installing a BCS is an attractive energy saving measure because of the additional benefits and opportunities it creates. A BCS optimizes the energy use of a building while simultaneously maintaining a comfortable indoor climate for the user.

Compared to no BCS, the indoor climate is improved which benefits the productivity and wellbeing of the users. BCS also reduces the cost for operating a building, by bringing down energy cost and cost for maintenance. The system can easily and quickly detect faults in the system or when a component needs replacement, which optimizes the maintenance of HVAC and other systems. It is also easier to operate the building as the BCS gives an overview of all systems and can provide easy and complete control over the various building operation systems. Other benefits are that BCS is rather easy and inexpensive to install, compared to other energy saving measures. BCS also create opportunities for intelligent control. If a BCS is present, it is easy to install new components or upgrade the existing technology to improve the system performance. The system can be programmed to recognize the user and adapt a room to the specific user's preferences.

Despite the benefits and opportunities, a BCS has to offer, there are also several challenges involved. In order for the BCS to function as designed, the interaction of both the building manager as well as the user with the system should be correct. This requires an understanding of how the system works and a well-designed user interface. If a user does not understand the system, they may feel that they no longer have freedom of choice to influence their environment. Users may interfere with the system by opening windows or changing settings. The complexity of the system requires that the building manager has a good understanding of how to operate and configure the system. If this is not done properly, the energy and cost savings are reduced and can even become higher than before. This may require technical training of a reliable and fast serviceability. Another challenge is the handling and storing of data. A huge amount of data is produced due to the constant monitoring of processes and indoor climate, which has to be stored in a safe way. Users may worry about the safety of sensitive information that is stored. Finally, there can be practical challenges related to installing the BCS. If components from different brands are used, they may not be compatible. However, the component selection from one brand may not cover the need for all the different components. Though it was said that a BCS is relatively easy to install, this is not always the case in existing buildings. If the existing building does not already have advanced HVAC systems, it can be expensive and time-consuming to install all the required system components and sensors. The benefits, opportunities and challenges from the perspective of the user, building manager and engineer are summarized in table 4.3.

4.6 Building automation control systems

Efficient control of the operational systems can be achieved by implementing a building automation control system (BACS) that automates the control of the functions. This is combined with technical building management (TBM) systems that inform the user about the operational energy management [23]. This results in the same benefits, opportunities and challenges as mentioned in table 4.4, though BACS is more advanced than BCS. The Standard EN 15232 lists the BACS and TBM functions that have a significant impact on the energy performance of buildings [24]. These functions are divided into the following categories: heating,

Table 4.3. Benefits, opportunities and challenges when using a BCS from user, building manager and engineer perspective.

	Benefits and opportunities	Challenges
User	Room adaptation to personal preferences Improved indoor comfort Increased productivity and wellbeing Improved user safety and security	Correct interaction with the system Users may feel they have no freedom of choice Users may interfere with the system, creating disturbances
Building manager	Reduced energy consumption through energy use optimization Reduced cost for operating and maintaining the building Improved fault detection Improved control of security systems Easy and complete control of various building operations	Safety of sensitive information Storage of massive amount of data Complexity of the system requires technical training of the building manager Will have the opposite effect if not installed and configured properly
Engineer	Relatively easy to install energy saving measurement Easy to upgrade and implement new technology	Compatibility of system elements from different brands Design of a user interface Complex to install advanced systems in existing building structures

cooling, ventilation, domestic hot water, lighting, blind control and data monitoring and diagnosis. The standard introduces four automation levels with corresponding efficiency factors for different building categories. These levels of automation are:

- D: Non energy efficient (no automation)
- C: Standard automation (reference value, standard BAC for new buildings)
- B: Advanced automation (advanced BAC with some TBM functions)
- A: High energy performance (high-performance BAC and TBM functions)

4.6.1 Energy savings from BACS

The efficiency factors can be used to estimate the savings that can be achieved when upgrading BACS by using equation (4.1). In this equation, E_{sav} is the energy savings in kWh, $E_{del,tot}$ is the total delivered energy, E_{th} is the thermal energy (heating, domestic hot water and cooling), E_{el} is the electrical energy (lighting and auxiliary) and η is the BACS efficiency factor for thermal and electrical energy. This method does not take into account optimization of the use of electrical devices and appliances to save energy. The efficiency factors, where C is assumed to be the standard level of the building, can be found in tables 4.4–4.6. It should be noted that the efficiency factors have to be adjusted so that the current automation level factor = 1 to correctly estimate the energy savings. The savings depend on the building typology, the total energy consumption of the building and the BACS installed.

$$E_{sav} = E_{del,tot} - \left[\left(\sum E_{th} \cdot \eta_{th}\right) + \left(\sum E_{el} \cdot \eta_{el}\right)\right] \text{ [kWh]} \tag{4.1}$$

To calculate the savings, it is necessary to know the thermal and electrical energy consumption of the building. The savings also depend on the thermal properties of the building envelope and the climate zone, as these highly influence the energy consumption and the effect of upgrading BACS decreases when the original delivered energy is lower [25–29]. It should be mentioned that the method proposed in

Table 4.4. BACS electrical energy efficiency factors for different building typologies taken from EN 15232 [24].

	Overall electrical factor				Electrical energy (lighting)				Electrical energy (auxiliary)			
	D	C	B	A	D	C	B	A	D	C	B	A
Residential buildings	1.08	1	0.93	0.92	–	–	–	–	–	–	–	–
Offices	1.10	1	0.93	0.87	1.1	1	0.85	0.72	1.15	1	0.86	0.72
Lecture halls	1.06	1	0.94	0.89	1.1	1	0.88	0.76	1.11	1	0.88	0.78
Educational buildings	1.07	1	0.93	0.86	1.1	1	0.88	0.76	1.12	1	0.87	0.74
Hospitals	1.05	1	0.98	0.96	1.2	1	1	1	1.10	1	0.98	0.96
Hotels	1.07	1	0.95	0.90	1.1	1	0.88	0.76	1.12	1	0.89	0.78
Restaurants	1.04	1	0.96	0.92	1.1	1	1	1	1.09	1	0.96	0.92
Commercial buildings	1.08	1	0.95	0.91	1.1	1	1	1	1.13	1	0.95	0.91

Table 4.5. BACS thermal energy efficiency factors for different building typologies taken from EN 15232 [24].

	Overall thermal energy				Thermal energy (heating)				Thermal energy (cooling)			
	D	C	B	A	D	C	B	A	D	C	B	A
Residential buildings	1.10	1	0.88	0.81	1.09	1	0.88	0.81	–	1	–	–
Offices	1.51	1	0.80	0.70	1.44	1	0.79	0.70	1.57	1	0.80	0.57
Lecture halls	1.24	1	0.75	0.50[a]	1.22	1	0.73	0.30[a]	1.32	1	0.94	0.64
Educational buildings	1.20	1	0.88	0.80	1.20	1	0.88	0.80	–	1	–	–
Hospitals	1.31	1	0.91	0.86	1.31	1	0.91	0.86	–	1	–	–
Hotels	1.31	1	0.85	0.68	1.17	1	0.85	0.61	1.76	1	0.79	0.76
Restaurants	1.23	1	0.77	0.68	1.21	1	0.76	0.69	1.39	1	0.94	0.60
Commercial buildings	1.56	1	0.73	0.60[a]	1.56	1	0.71	0.46[a]	1.59	1	0.85	0.55

[a] These values depend greatly on the heating and cooling demand for ventilation.

Table 4.6. Domestic hot water energy efficiency factors for different building typologies taken from EN 15232 [24].

	Domestic hot water			
	D	C	B	A
All buildings	1.11	1	0.90	0.80

EN 15232 is only an estimation of the energy savings that can be achieved, and it can underestimate the actual savings [25]. For a more precise result, the potential energy savings should be investigated by using simulation tools. The effect of BACS on improving the indoor climate of the building can also be assessed using simulation tools. However, advanced BACS is difficult and time-consuming to model in energy performance simulations and is only useful in the case of research. Studies estimated savings on the delivered energy between 8–21% for a single-family house in Norway when upgraded from D to a higher level [25]. For apartment blocks in Finland, space heating demand savings of up to 60% can be achieved with intelligent control, though if the system cannot respond to occupant behaviours (in this case: opening windows) it will result in an increased space heating demand [30]. In offices in different European cities, 5–45% can be saved when specific control functions are upgraded [29].

4.7 Energy flexibility

The energy market has to be able to supply an amount of energy that is more or less equal to the demand and it should have a capacity that is higher than the peak load of the system. Renewable energies can often not meet peak demands meaning this energy is produced from non-renewables resources, like coal and oil. The constantly increasing energy demand puts a strain on the energy production and the environment because of this. The capacity of the grid has to be enlarged to provide for the high peak demands, which occur in short intervals only. This is a costly investment for grid companies and results in higher energy prices for the consumer. In order to reduce the energy cost and the price variation during the day, the consumers have to be more active in how and when they consume energy [31].

These challenges require a flexible energy system. A building (automation) control system can help to reduce the energy costs by shifting the peak loads and adapting the energy demand to the hourly grid tariff. It can also evaluate when it is best to store excess energy and when to use stored energy instead of energy from the grid. Peak load shifting is related to the level of agreement between the demand and the supply of energy. With peak load shifting the peak loads are moved to periods where the demand is low. This reduces the total peak power of the system and therefore minimizes the required capacity of the energy system. If it is not possible to shift the loads, the demand from the grid can be reduced by using stored energy during peak hours as supply. Through this, the pressure on the grid and the use of fossil fuels is decreased. In this way, the existing infrastructure may not need to be expanded, which results in significant cost savings for both the producer and the consumer [31].

4.8 User interaction

Users influence the efficiency of a BCS with their actions. In order to reduce this disturbance factor, the system should be able to interact with the user, by providing insight in the energy consumption and indoor climate and respond to the user preferences. Feedback strategies can be implemented to increase the knowledge and awareness of the user so that the user behaviour is improved, and energy savings are achieved [32–34]. However, feedback does not necessarily result in decreased energy use or long-term changes in user behaviour [35–37].

To ensure that the BCS functions as intended and the disturbances from user actions are minimized, it should be easy to interact with the system and the system should take into account the user's preferences and needs [37]. The user should also be motivated, whether financially or for other reasons, to interact with the system [38]. If the user is not motivated or interested, he may try to adapt the environment, override the system or switch it off. In this way, no energy savings are achieved at all.

4.9 Future benefits and challenges

The need for energy management systems is therefore driven by the option of effectively handling energy streams through the use of smart controls and a supervisory scheme consisting of communicating with both loads and generation of energy in order to manage demand and supply, and the option of communicating with the internal network in order to schedule the amounts of production needed to benefit economically with exchanging energy to the grid [39]. The integration of the forecasting systems, which can forecast hourly power generation according to the weather forecasting inputs, could create an ideal timetable so that economically optimized energy delivery can be maintained to meet load demand. Some trials indicate that the weekly cost would be reduced by 28%. Further study could be carried out to enhance this technique by incorporating more appropriate variables such as a town or region's agricultural and business profiles [40].

Peak demand tax is one of the main parts of the power bill of a customer and is the calculation of the quantity of energy needed during peak demand phases. To meet peak demand, utilities are required to purchase reserves that remain untapped most of the time and are to be used only for short periods of high demand or other reserves failure. In addition, smart buildings are generally equipped with technologies for generating renewable energy. However, Local production of energy from renewable sources and energy demands are unfortunately misaligned. Therefore, an intelligent technology could be required to alter indoor climate circumstances in order to change energy requirements from peak to off-peak hours energy demands. In addition, if the building is equipped with electrical energy storage, renewable energy can be used to charge the storage when the cost of electricity is low, and the storage can be discharged during high-cost periods. Therefore, using storage in combination with renewable energy sources could assist in optimising a smart building's cost-effectiveness. For intermittent sources (e.g. solar and wind), a time shift in renewable energy is especially valuable [41].

Future research should continue to develop algorithms and study individual practices to enhance the energy management strategy's effectiveness and flexibility. The prospective advantage of transmitting energy use to families in a structural sense (i.e. viewing housing as a portion of a smart grid), was also studied and acknowledged as an important prospective vehicle for deploying energy management alternatives in households [42]. It has been shown that the impact of feedback on end-users is more efficient the more immediate it is. In addition, information depiction should be customized to suit the requirements, practices and education of end-users. It has also been shown to be a benefit for consumers to be able to compare their building energy use with benchmarks as well as their own historical information. What is an appropriate comparative, however, will also differ between customers. For example, some users may be concerned in linking their own use with others use, thus providing a comparison aspect, while others may not be involved in engaging in this sense at all. In addition, people can be categorized according to distinct guidelines (e.g. by buildings of comparable magnitude), buildings with similar occupancy times or structures with similar median uses of energy [42]. The most difficult element of user motive is that whatever sort of feedback is used, to satisfy the domestication requirement, the technology needs to continuously evolve. The method involves end-users adapting the technology to their own requirements and aspirations. However, this method has an adverse side effect from a visualization point of view [43], as any positive impact of any visualization techniques used for the advantage of end-users ultimately stop as part of the domestication framework. Consequently, any leadership scheme, regardless of how efficient it may be, will only stay efficient if information display is versatile and adjustable over time. The difficulty resides in not only providing appropriate information to end users but also continuous participation in this method of adjustment and customization.

Recently, gamification opportunities (i.e. the use of videogame components and non-gaming ideas to enhance user experience and interface involvement), has begun to be explored extensively. A future gamification component may be included if you have the choice to match your own energy use with that of others. Finally, to understand a correct home automation scheme capable of interacting with customers, another important obstacle, depicted by the awareness of customers about smart technologies, must be drawn into account. Not only do smart devices need to be properly fitted and equipped, but the user's understanding of smart devices as well as their energy management awareness need to be steadily enhanced.

References

[1] Gaddam A, Mukhopadhyay S C and Sen G G 2010 Elder care based on cognitive sensor network *IEEE Sens. J.* **11** 574–81

[2] Mileo A, Merico D and Bisiani R 2008 Wireless sensor networks supporting context-aware reasoning in assisted living *Proc. of the 1st Int. Conf. on Pervasive Technologies Related to Assistive Environments* (ACM), p 54

[3] IPCC 2007 Intergovernmental panel on climate change *Clim. Chang. 2007 Synth. Rep.*
[4] Gellings C W 2009 *The Smart Grid: Enabling Energy Efficiency and Demand Response* (Lilburn, GA: The Fairmont Press)
[5] Lobaccaro G, Carlucci S and Löfström E 2016 A review of systems and technologies for smart homes and smart grids *Energies* **9** 348
[6] Gungor V C *et al* 2011 Smart grid technologies: communication technologies and standards *IEEE Trans. Ind. informatics* **7** 529–39
[7] Berlad A L, Salzano F J and Batey J 1976 On enthalpy management in small buildings *Energy* **1** 429–43
[8] Aman S, Simmhan Y and Prasanna V K 2013 Energy management systems: state of the art and emerging trends *IEEE Commun. Mag.* **51** 114–9
[9] Erol-Kantarci M and Mouftah H T 2011 Wireless sensor networks for cost-efficient residential energy management in the smart grid *IEEE Trans. Smart Grid.* **2** 314–25
[10] Beaudin M and Zareipour H 2015 Home energy management systems: a review of modelling and complexity *Renew. Sustain. Energy Rev.* **45** 318–35
[11] Rao P B and Uma S K 2015 Raspberry Pi home automation with wireless sensors using smart phone *Int. J. Comput. Sci. Mob. Comput.* **4** 797–803
[12] Robles R J and Kim T 2010 Applications, systems and methods in smart home technology: A. *Int. J. Adv. Sci. Technol.* **15** 34–48
[13] Carlucci S, Causone F, De Rosa F and Pagliano L 2015 A review of indices for assessing visual comfort with a view to their use in optimization processes to support building integrated design *Renew. Sustain. Energy Rev.* **47** 1016–33
[14] Carlucci S and Pagliano L 2012 A review of indices for the long-term evaluation of the general thermal comfort conditions in buildings *Energy Build.* **53** 194–205
[15] Loveday D L, Virk G S, Cheung J Y M and Azzi D 1997 Intelligence in buildings: the potential of advanced modelling *Autom. Constr.* **6** 447–61
[16] Preiser W F E and Schramm U 2002 Intelligent office building performance evaluation *Facilities* **20** 279–87
[17] Robathan D P 1994 *The Future of Intelligent Buildings* (London: Alfred Waller in association with Unicom), pp 259–65
[18] Arkin H and Paciuk M 1997 Evaluating intelligent buildings according to level of service systems integration *Autom. Constr.* **6** 471–9
[19] Boyd D 1994 *Intelligent Buildings and Management* (London: Alfred Waller in association with Unicom), pp 7–18
[20] Yang J and Peng H 2001 Decision support to the application of intelligent building technologies *Renew. Energy* **22** 67–77
[21] Dounis A I and Caraiscos C 2009 Advanced control systems engineering for energy and comfort management in a building environment—a review *Renew. Sustain. Energy Rev.* **13** 1246–61
[22] Abel E *et al* 2003 Achieving the desired indoor climate-energy efficiency aspects of system design. Studentlitteratur
[23] Felius L, Dessen F and Hrynyszyn B 2019 Retrofitting towards energy efficient homes in European cold climates—a review *Energy Effic.* **13** 101–25
[24] European Committee for Standardization 2017 EN 15232-1:2017. Energy performance of buildings — Part 1: impact of building automation, controls and building management

[25] Felius L, Hamdy M, Hrynyszyn B and Dessen F 2020 The impact of building automation control systems as retrofitting measures on the energy efficiency of a typical Norwegian single-family house *IOP Conf. Ser.: Earth Environ. Sci.* **410** 012054

[26] Vallati A, Grignaffini S, Romagna M and Mauri L 2016 Effects of different building automation systems on the energy consumption for three thermal insulation values of the building envelope *2016 IEEE 16th Int. Conf. on Environment and Electrical Engineering (EEEIC)* (IEEE), pp 1–5

[27] Ippolito M G, Sanseverino E R and Zizzo G 2014 Impact of building automation control systems and technical building management systems on the energy performance class of residential buildings: an Italian case study *Energy Build.* **69** 33–40

[28] Sanseverino E R, Zizzo G and La Cascia D 2013 Economic impact of BACS and TBM systems on residential buildings *2013 Int. Conf. on Clean Electrical Power (ICCEP)* (IEEE), pp 591–5

[29] Gyalistras D, Gwerder M, Oldewurtle F, Jones C and Morari M 2010 Analysis of energy savings potentials for integrated room automation *Clima-RHEVA World Congress*

[30] Reda F, Pasini D, Laitinen A and Vesanen T 2019 ICT intelligent support solutions toward the reduction of heating demand in cold and mild European climate conditions *Energy Effic.* **12** 1443–71

[31] Karlsen S S 2018 Investigation of grid rent business models as incentive for demand-side management in buildings – a case study on fully electric operated houses in Norway *NTNU*

[32] Ouyang J and Hokao K 2009 Energy-saving potential by improving occupants' behavior in urban residential sector in Hangzhou City, China *Energy Build.* **41** 711–20

[33] Fischer C 2008 Feedback on household electricity consumption: a tool for saving energy? *Energy Effic.* **1** 79–104

[34] Grønhøj A and Thøgersen J 2011 Feedback on household electricity consumption: learning and social influence processes *Int. J. Consum. Stud.* **35** 138–45

[35] Abrahamse W, Steg L, Vlek C and Rothengatter T 2005 A review of intervention studies aimed at household energy conservation *J. Environ. Psychol.* **25** 273–91

[36] Hargreaves T, Nye M and Burgess J 2013 Keeping energy visible? Exploring how householders interact with feedback from smart energy monitors in the longer term *Energy Policy* **52** 126–34

[37] Nilsson A, Bergstad C J, Thuvander L, Andersson D, Andersson K and Meiling P 2014 Effects of continuous feedback on households' electricity consumption: potentials and barriers *Appl. Energy* **122** 17–23

[38] Buchanan K, Russo R and Anderson B 2015 The question of energy reduction: the problem (s) with feedback *Energy Policy* **77** 89–96

[39] Graditi G, Di Silvestre M L, Gallea R and Sanseverino E R 2014 Heuristic-based shiftable loads optimal management in smart micro-grids *IEEE Trans. Ind. Informatics* **11** 271–80

[40] Chen C, Duan S, Cai T, Liu B and Hu G 2011 Smart energy management system for optimal microgrid economic operation *IET Renew. Power Gener.* **5** 258–67

[41] Graditi G, Ippolito M G, Telaretti E and Zizzo G 2016 Technical and economical assessment of distributed electrochemical storages for load shifting applications: an Italian case study *Renew. Sustain. Energy Rev.* **57** 515–23

[42] Löfström E 2014 Smart meters and people using the grid: exploring the potential benefits of AMR-technology *Energy Procedia.* **58** 65–72

[43] Hirsch E and Silverstone R 2003 *Consuming Technologies: Media and Information in Domestic Spaces* (Routledge)

Chapter 5

Energy production in buildings

Bjørnar Hamre, Tania Bracchi, Laurina C Felius, Odne S Burheim, Bruno G Pollet and Jacob J Lamb

Smart buildings plan their power usage and storage. While passive houses have only a limited amount of net energy use, energy positive houses deliver energy to their surroundings. Smart energy houses produce energy renewably when sources are available and store it for later use. They can also use the weather forecast to predict their energy requirement so they can buy energy during the hours when it is cheaper. This chapter will detail renewable energy production from wind and solar for use in buildings.

5.1 Introduction

There are in essence two technological branches relevant for electrical energy production in buildings; photovoltaic (PV) cells and wind turbines. This chapter will focus on these branches with regard to electrical energy generation. Furthermore, the technologies most relevant for utilisation in smart buildings will be discussed, along with an assessment of the cost of the different technologies. When considering technology costs, it must be noted that these values are indicative only, as life expectancy, exchange rates, dealer mark-up and technology development will cause the price to rapidly change.

5.2 Solar electrical energy

Solar radiation is abundant, dominant, and in many respects the main long-term sustainable energy option for all human activity. PV cells are a technology that can convert this solar energy directly into electrical energy. Photons emitted from the Sun enter the PV cell where it, depending on its energy, can excite an electron. Because PVs are semiconductors made to act like a diode, excited electrons are forced through an external circuit, driving a current.

The efficiency of extracting usable energy from the resource in PV technologies is low compared to other technologies. Most PV cells convert under 20% of the energy

doi:10.1088/978-0-7503-3259-0ch5

radiated onto the cell into electrical energy. However, this energy resource is completely renewable and available almost everywhere on the surface of the planet. Therefore, PV technology is one of the most promising technologies of the future, and as a result the sales and production of PV systems have skyrocketed in recent years. Especially since China entered the PV market, the industry has hugely expanded. The technology development continues, improving the efficiency of the systems, reducing their shelf price and strengthening their position as a sustainable electricity producer.

5.2.1 Solar production

The production of a PV system depends on the geographical conditions and the efficiency of the components in the PV system. Expected yearly production from a PV system can then be calculated using the following equation.

$$E = PR\eta AI \tag{5.1}$$

here, E is the energy produced in kWh per year, PR is the performance ratio of the PV system, η is the module's efficiency, A is the module area in m^2 and I is the yearly irradiance on the module surface in kWh m^{-2} $year^{-1}$. The efficiency of a PV cell is calculated by dividing the maximum power produced from the cell (W) by the power radiated onto the cell surface by the Sun (solar radiance in W m^{-2}) multiplied by the cell area in m^2. PV systems are most efficient when the modules are placed at an angle perpendicular to the incoming sun rays. A guideline for optimal orientation of the panel is to subtract the geographical latitude from 90°.

5.2.2 Types of PV panels

The most relevant PV panels for a residential energy production system are thin-films, multicrystalline silicon (multi-Si) and monocrystalline silicone PVs (mono-Si). These are elaborated in detail here.

5.2.2.1 Thin-film and amorphous PV panels

Thin-film PV cells are considerably thinner than their alternatives. They can have a thickness of only a few nanometers for the active materials [1]. Because of their dimensions, these PV cells are mechanically supported by another material like glass or plastic (laminate). Thin-film modules can be designed to be more flexible, and are, therefore, easily integrated into building architecture. The small thickness leads to less material use and as a result a smaller production cost. Common materials to use in thin-film PVs are cadmium telluride (CdTe), copper indium gallium selenide (CIGS) and amorphous silicon[1] (a-Si). Currently, CdTe PVs have the largest market share of the three. All thin-film technologies together amounted to about 6% of the total production in 2016 [2].

Both CdTe and CIGS PVs have reported record efficiencies of over 20% [3]; however, the materials used in these PVs raise issues that effects their cost and

[1] Amorphous silicon is silicon with a non-crystalline structure.

demand. They both contain cadmium, a toxic element that has mostly been phased out in other industries. Despite this, recycling systems have been set up for commercial CdTe modules to avoid the potential threat of cadmium emissions, and also because of the scarcity of tellurium. Indium is a key element in CIGS, and its scarcity is a concern for scaling up the CIGS PV production [1]. CIGS PV production actually declined in 2017 after years of negative operating metrics [4].

A-Si PV modules are often bypassed as the first choice for a PV system because of their low efficiency. The record efficiency of an a-Si PV is 10,2% [3]. An increased area of active PV cells is needed for PVs with lower efficiency to reach the same power output. Efficiency is an important factor of the PV cost reduction due to the extra soft costs of installing a larger PV system for a wanted output. The soft costs in regard to PVs are for example the costs of installing labour, sales tax and sales profit. Therefore, a-Si technology has had a slower development than other PVs and has a reducing market share [4]. This makes CdTe the only thin-film technology with a positive outlook in 2017.

5.2.2.2 Crystalline silicon PV panels

Both multi-Si and mono-Si are made from crystalline silicon. Together they accounted for 93% of the total PV module production in 2016 [2]. Silicon is a common semiconducting material found in large quantities in the Earth's crust. Most PV cells are 10 × 10 cm or 15,6 × 15,6 cm. Each cell has an output voltage of about 0.5 V. Several cells are coupled together to make a PV module. The voltage is boosted by coupling the cells in a series, and the current output is boosted by coupling the PV cells in parallel. Each module is coated with a material to protect the cells from mechanical erosion. This keeps the electrical contacts on the cell from degrading from exposure to moisture.

5.2.2.3 Multicrystalline silicon PV panels

The multi-Si PV is the most common PV cell. About 70% of the total crystalline silicon production was based on multi-Si in 2016 [2]. The cell consists of multiple silicon crystals, hence the name. When molten silicon is cooled in a cast, it solidifies into directional crystals, making a solid block of multi-Si, later cut into ingots. The thin wafers recognized from solar panels are cut from the ingots. There is some electrical resistance in the transition between crystals, lowering the electronic quality of the silicon, resulting in a voltage drop causing an efficiency drop. Present efficiency for new commercial multi-Si PV modules averages at around 16% [2]. The record efficiency of the multi-Si PV is currently 20.8% [3]. Much research is invested in developing better multi-Si cells, resulting in the average efficiency rising each year. Expanding production also reduces the price of multi-Si modules.

5.2.2.4 Monocrystalline silicon PV

As hinted in the name, a mono-Si PV cell is contrived of one single crystal. It is the most efficient of the silicon PVs because of the purity of the material. The record efficiency for a mono-Si cell is 26.7% [3]; however, the production of mono-Si has a higher economic cost and is more time consuming than for multi-Si. The efficiency of the module in the PV system is crucial for optimal production and maximizing profits. Therefore, mono-Si is regarded as a competitive PV technology despite a higher production cost.

5.2.3 Inverters for PV systems

The power output from the PV module is in the form of a DC current. To make use of this current, one or several inverters are added to the PV system. The inverters manipulate the current from the modules to make it AC and, therefore, usable with the power system of the house. Some energy is lost in this process. It is possible to use either a string inverter or micro-inverters (one small inverter for each panel) for the system. String inverters have higher efficiency (up to 98%), than micro-inverters (90%–95%). However, micro-inverters have other advantages. In a PV system with a string inverter, all modules are limited to produce at the same voltage because they are connected to the same centralized inverter. As such, if one PV module has a reduced power output, the rest of the system will be reduced to that level. For example, when there is shading on a part of the PV system, the whole system is affected. On the other hand, micro-inverters handle one PV module each. Therefore, the production from one module is unaffected by the production from the other modules. Depending on the local conditions, the performance ratio of the system can be better with the less efficient micro-inverters than with one string inverter. Micro-inverters can be more difficult to replace and are more costly than string inverters, but they have a longer life expectancy.

5.2.4 Energy payback time

Energy payback time (EPBT) is the time the PV system needs to be operational to make up for the energy used to produce the PV system. Therefore, EPBT is the energy used to produce the component(s) of the PV system divided by the energy produced each year by the PV system. The EPBT is proportional to local irradiance. Depending on technology and location of the PV system, the total EPBT ranges from 0.7 to over 2 years. Numbers from 2011 point to silicon extraction and wafer production as the biggest contributors to EPBT. This is increased for the mono-Si PV systems that are produced through time consuming and energy demanding processes, resulting in a much longer EPBT. For irradiance at 1000 kWh per m^2 per year, the mono-Si module production contributes to about 80% of the system EPBT. For multi-Si, the number is 70%. The inverter contributes with about 0.25 years to the EPBT of a PV system. For the CdTe PV system, the module (mostly laminate) is the biggest contributor to EPBT. The total EPBT of the CdTe PV system is the smallest of the three [2].

5.2.5 PV costs

China entered PV production industry in the mid-2000s. They rapidly expanded their production to become the largest PV producing country in the world, significantly lowering the global prices of PV technology in the process. The PV modules themselves have seen extensive decreases in price as a result. Over 70% of the hardware cost was credited to the PV module in the early 2000s; however, by 2016, it was under 50% of the hardware cost [2]. In addition to the hardware cost for the PV system, there are soft costs. The soft costs for residential PV systems have decreased at a slower rate than the hardware cost. Because of this, the soft cost percentage of the total PV system cost has increased from 50% in 2010, to 68% in 2017 [5]. It is difficult to use a global scale for the total cost of a residential PV system, as markets vary considerably with incentives, labour costs and demand. Two large solar markets are USA and Germany, and in 2016 they had an average price for residential PV systems of 1.25 and 1.5 USD per W, respectively [2, 5].

The financial payback time (PBT) for the PV system is here calculated as the cost of the PV system (excluding soft costs), divided by the cost of the energy saved each year (operational hours multiplied by the average electricity price). For example, if Southern Europe is considered, a PV system would probably have around 1700 operational hours per year [6]. If we then use an arbitrary electricity price of 0.35 USD per kWh^{-1}, the PBT can be calculated. Using these values, the PBT would be 2.5 years. The PBT is highly affected by the price of electricity, for example, if the cost of electricity was lower at a value of 0.12 USD per kWh^{-1}, the PBT would become close to 6 years. Furthermore, it is also affected by local and national incentives, investment loan costs and the PV system soft costs.

5.2.6 Comparison of different PV technologies

In table 5.1, an overview of PV technologies is presented. Comparing the different technologies is important when choosing the best possible system. Especially noticeable in the table is the large difference in PBT for Germany and USA. The contrast is large because of the large different in prices of electricity used in the calculation for the different countries.

Table 5.1. Efficiencies and costs for different PV technologies [2, 5–10].

Component type	Efficiency (%)	Life exp. (y)	Cost ex. soft ($/W)	EPBT (y)	PBT ex. soft (Y)
Module					
CdTe	16	30	0.28	0.54	—
Mono-Si	17	30	0.371	1.2	—
Multi-Si	16	30	0.359	1.1	—
Inverter					
String	98	10	0.08–0.22	0.25	—
Micro	90–95	25	0.37	0.25	—
Full system					
Germany	80–90	25	1.5	0.7–2	2.5
USA	80–90	25	1.25	0.7–2	6.0

5.2.7 Challenges for PV systems

The demand for PV systems largely depends on the profit they can provide. This is a challenge for all new technologies. A new technology is often opted out of for a more mature technology with better documented results. Without incentives or payment schemes, the investment cost of a PV system can seem too large for most residential buildings, even though it will turn a profit over its lifetime. Electricity prices are also a large factor when calculating savings from a PV system. Low electricity prices from cheap energy sources can make installing a PV system economically unfavourable. As the profit of an investment often is the major driving factor of demand, this can become a large challenge for PVs.

For PVs to produce energy, the PV modules need exposure to the Sun. If local conditions prevent sun exposure, the production potential will drop. The further the location from the Equator, the lower the Sun is in the sky. This causes shading of the PV systems, which occurs more readily with a low hanging sun. In addition, the daylight hours are lower (or non-existent) in the winter in areas far from the Equator, and snow covering the PV systems adds to the reduction in solar irradiance on the PV systems. Squeezing a profit from a PV system that is operational only during summer, when the price of electricity often is low, is an impossible project. Furthermore, with respect to environmental impact, in many cases the installation of PV systems far from the equator has a net increase in greenhouse gas (GHG) emissions if the local grid is already supplied with renewable energy (e.g. hydroelectric power).

5.2.8 Conclusions and future development

PV technology for harvesting solar energy is advancing rapidly, as the demand for renewable and off-grid energy sources increases. In most cases, installing a PV system on a residential building will have an economic profit over the lifetime of the system; however, the state of the market plays a large role in the availability and pricing of the PV system. For countries who ventured into the possibility of decentralized residential PV systems with incentives and specialized loans, the demand grew rapidly. The increasing rate of building utility scale PV power plants has driven the prices of PV models down, significantly lowering the installation costs of PV systems. Coupling this to the increasing efficiency of PV modules, a further rise in both demand and availability for PV systems is predicted.

5.3 Wind electrical energy

The combined renewable power capacity in 2018 was 2378 GW of which 591 GW was from wind energy through wind turbines on an industrial scale [11]. From 2014 the worldwide wind power capacity grew nearly 60% in 4 years. Although large-scale wind-energy production has taken form, its sustainable growth has some issues. For instance, the primary obstacles of large-scale onshore wind farms are available locations, the effect of performance on the grid power, government acceptance and losses in electricity transmission and allocation to consumers [12]. Small wind turbines (SWTs) are an option that reduces some of these obstacles.

5.3.1 The urban environment

Megawatt-scale wind turbines are usually connected to the grid or to power industrial plants and commercial activities; however, in urban areas SWTs could be integrated directly onto buildings. With reference to the objective of smart buildings, efficient use of spaces and local production of energy is required. Despite this, the conditions in the urban environment are significantly different to open areas, with the average wind speed being in general lower and the turbulence higher.

The urban wind system is defined by small annual mean wind speeds (AMWS) and a more turbulent stream in the atmospheric boundary layer (ABL) as compared to rural regions' open terrain. This is due to the quickly altering wind direction and the presence of human-made obstacles. Moreover, the low AMWS is the consequence of irregular land topography and obstacles, whereas the enhanced turbulent flow is the consequence of wind interacting with structures and obstacles.

5.3.2 Atmospheric boundary layer

The wind speed distribution over the Earth's surface is defined as the atmospheric boundary layer (ABL), and it is usually described by the log-law or power-law profile models function of surface roughness. Due to the presence of obstacles (e.g. buildings), the ABL obtains a much more complex pattern (figure 5.1).

Despite the average wind speed in urban areas being generally lower than in open areas, an interesting phenomenon is observed on the roof of the buildings (i.e. flow speeds-up). This increase can be up to around 20% of the undisturbed wind in front of the building, depending on the wind direction with respect to the building [13]. This could suggest the implementation of wind turbines on the roof of a building, without special modification of the building itself. However, one precaution is that the wind turbine should not be placed in the wake created by the flow separation on the roof of the building.

5.3.3 Wind turbines in the urban environment

The local wind direction is expected to vary with a time scale, which is a function of the height of the building and the wind speed [13]. A rough estimation of the wind turbine dimensions, based on the stream tube analysis for viscous flows, suggests that in order to avoid variation in structural loads due to turbulence, the maximum diameter of the turbine should be around 20% of the height of the building (i.e. for a building 30 m tall, the wind turbine diameter should not exceed 6 m). Therefore, the dimensions of a wind turbine in an urban environment are limited compared to rural wind turbines. This is partly due to the fact that the turbulence level is much higher around buildings, which decreases the production and increases the loads on the turbine structure. Furthermore, the wind speed on the top of a building can reach up to 20% higher than the wind speed far in front of the building [13]. Therefore, the best position for a wind turbine in an urban area is over the highest building, where

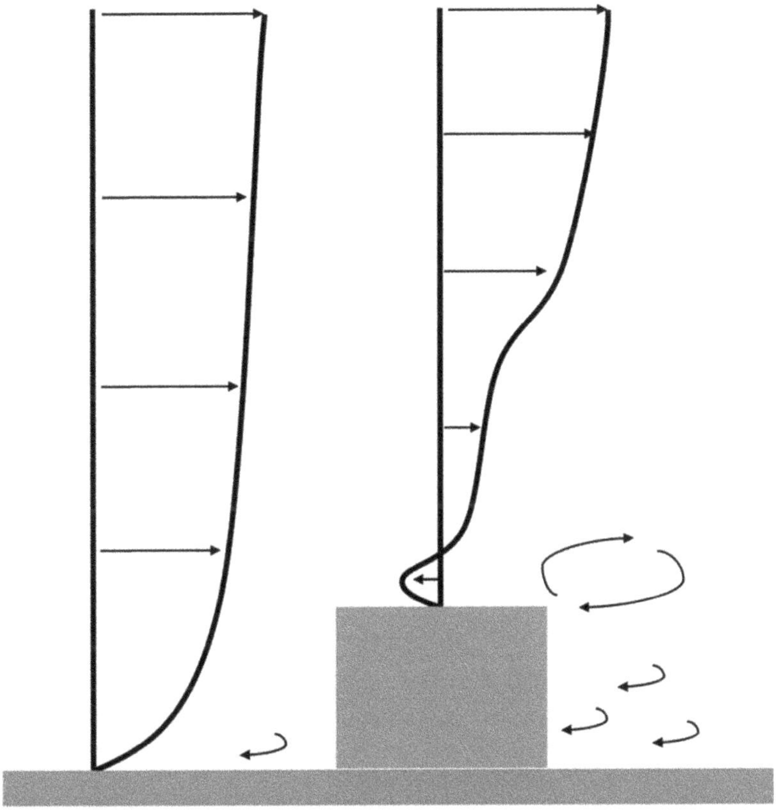

Figure 5.1. The effect of buildings on oncoming wind profiles.

the wind speed is higher, and the turbulence is lower. Nevertheless, the feasibility of wind energy as a resource in urban wind should be investigated further [14].

5.3.4 Vertical axis versus horizontal axis wind turbines

The promising opportunities for installing SWTs in the urban landscape have been shown by recent developments in wind energy technology and studies on turbine design and wind characteristics. Either vertical-axis wind turbines (VAWTs) or horizontal-axis wind turbines (HAWTs) may be installed in such areas. The VAWTs may have either a lift- or drag-driven rotor. The most suitable type of turbine in the urban landscape depends on cut-in wind speed, flexibility in installation and operation, height-limit and architectural integration with the existing morphology. For example, wind flow monitoring is required in HAWTs, while the set spin axis leaves VAWTs more visually attractive. With reference to large scale wind turbines, the most successful is the HAWT with three blades. The nominal power of such commercial wind turbines today is around 4 MW (110 m diameter onshore turbine), and 6 MW (150 m diameter offshore turbine).

Table 5.2. Comparison of technical specifications of wind turbines.

	VAWT	HAWT
Rated power (kW)	500	500
Swept area (m^2)	955	1370
Rated wind speed ($m\ s^{-1}$)	12.5	15
No. of blades	2	3
Tower height (m)	50	47
Turbine diameter (m)	34	42
Blade length (m)	54.5	18
Blade material	Aluminium	Glass fibre
Yaw mechanism	No	Yes
Pitch/active stall mechanism	No	Yes
Gear box	Yes	Yes
Support wires	Yes	No
Generator position	Ground	On tower
Rotation speed	Semi-variable, 28–38	Constant 19/28
Total weight (kg)	72,200	68,000

The HAWT is more advanced and established in the wind industry when compared with the VAWT and usually performs well in residential environments or accessible fields where wind stream can be comparatively constant and unidirectional. However, VAWT often outperforms HAWT in the built environment and especially on rooftops with frequently turbulent flow due to its reduced sensitivity to changes in wind direction and its ability to generate electricity even under complicated and variable wind circumstances. In addition, by adding a diffuser-shaped shroud to the VAWT design, it is possible to take greater advantage of the high wind speeds created by accelerated flow on the rooftops of buildings, thereby increasing power generation [15].

HAWTs' other disadvantages are the elevated tip noise that can impact rural communities and their low production from low wind speed of less than 6 $m\ s^{-1}$, with a cut-out rate of around 25 $m\ s^{-1}$ [16]. VAWTs have reduced visual impact and produce less noise, particularly the silent Savonius scheme, which can generate energy in winds as low as 1 $m\ s^{-1}$, to around 65 $m\ s^{-1}$ at cut-out velocity. However, some studies [16] show that the Darrieus vertical-axis wind turbine shows a better behaviour in turbulence condition and skewed flow. A common concern about the Darrieus vertical-axis turbine, and in general lift-driven VAWTs, is that they require an additional aid to start. Despite this, studies have shown that with special attention in the design phase of the blades, a Darrieus vertical-axis turbine has the capability to self-start at low rotational speeds [17]. Table 5.2 gives a comprehensive comparison between a VAWT and a HAWT wind turbine.

5.3.5 Wind turbine performance

To determine the potential electrical production of a wind turbine in a specific area, some information about the local wind environment is required. The production can then be calculated for a specific wind turbine proportional to the cube of the wind speed.

$$P = \frac{1}{2}\rho U^3 S C_P \quad (5.2)$$

here, ρ is the air density, U is the wind speed, S is the swept area and C_P is the power coefficient. Note that the swept area for HAWTs is the rotor area and for VAWTs is the frontal area of the wind turbine (turbine diameter multiplied by rotor blade height).

The aerodynamic efficiency of the wind turbine is represented by the power coefficient. It is a function of the tip speed ratio, which can be calculated by:

$$\lambda = \frac{\omega R}{U} \quad (5.3)$$

where ω is the rotational frequency of the turbine, R is the turbine radius and U is the wind speed.

5.3.6 Construction standards

Wind turbines must adhere to the IEC 61400 international standard to be classed as safe, durable and high performance with the ability to withstand high winds, turbulence and extreme weather events. Furthermore, the IEC 61400–2 standard is used to define the regulations for SWT with a swept area below 200 m^2. In general, SWTs have a smaller rotor size as they are designed to not interfere extensively with the vertical wind profile. Moreover, many local governments have their own amendments to the size of SWTs to reduce their impact (e.g. visual impact) on the local environment. The size of a wind turbine is defined based on the application, but generally the capacity of 100 kW divides small from large wind turbines.

5.3.7 Cost assessment

The cost of small wind turbines is highly dependent on the producer, quality of the components and the installation cost. In addition, the maintenance costs depend on the environmental condition of the site where the wind turbine is placed. Table 5.3 gives an overview of the power production and costs of three different wind turbines (calculations based on the technical information for each turbine).

The cost of a small wind turbine ranges between 3 and 6 USD per W. The levelized cost of energy (LCOE) is expressed in USD per kWh and is a common index for comparison of investment and performance for energy generation technologies. LCOE is defined as:

Table 5.3. Energy production and cost of urban wind turbines.

	UGE 4K GT Darrieus Turb. 4 kW - $24,500		Helix Wind s594 Savonius Turb. 4.5 kW - $16,500		Kingspan KW3 HAWT3 kW - $17,000	
Average annual wind speed	Produced energy (kWh/y)	AEO ($/kWh/y)	Produced energy (kWh/y)	AEO ($/kWh/y)	Produced energy (kWh/y)	AEO ($/kWh/y)
4	3000	8.17	45	366.67	3200	5.31
5	5000	4.90	495	33.33	6000	2.83
8	13 000	1.88	3510	4.70	11 200	1.52
10	16 000	1.53	6500	2.54	12 000	1.42
12	16 000	1.53	9495	1.74	N.A.	N.A.

$$\text{LCOE} = \frac{\text{Total costs over lifetime}}{\text{Electricity produced over lifetime}} \tag{5.4}$$

LCOE for a wind turbine depends on the initial investment, the average wind speed and the calculated lifetime. The lifetime of wind turbines varies with the model and the environmental conditions on-site, but it is usually between 15 and 30 years. The following LCOE values are calculated over a 15 year lifetime [18]:

Average wind speed of 5 m s^{-1} and 3 USD per W: 0.2 USD per kWh
Average wind speed of 5 m s^{-1} and 6 USD per W: 0.4 USD per kWh
Average wind speed of 10 m s^{-1} and 3 USD per W: 0.08 USD per kWh
Average wind speed of 10 m s^{-1} and 6 USD per W: 0.15 USD per kWh

5.3.8 Wind turbine noise pollution

Wind turbine noise has two primary causes; aerodynamic noise from the blades and mechanical noise from the components of the drive train. With growing turbine blade tip velocity, aerodynamic noise increases. A VAWT generally has a tip velocity about half of a HAWT's tip velocity and therefore less aerodynamic noise is produced. Since a VAWT has ground-level drive train components, the potential noise from these components will not propagate as easily as when the drive train components are located on top of a tower. The Darrieus turbine is rotating slower than the same length HAWT; therefore, it will generate less noise than the HAWT.

5.3.9 Challenges for urban wind turbines

The development of large-scale wind turbines has continued at a rapid pace in recent years, yet SWTs still require further development. The main problem is the integration of the wind turbine on the building, as the location can have significant influence on the performance, as well as the noise pollution. Furthermore, the cost of SWTs is much higher than that of large wind turbines per unit of power produced.

Although SWTs have a lower operational cost in many cases, the installation costs can be significantly higher than large wind turbines per unit of power produced. Therefore, SWTs struggle to compete with the cost benefits of larger wind turbines. This could be elevated with government-sponsored schemes to increase the implementation of SWTs into the urban environment.

5.3.10 Conclusions and future developments

There are many instances where governments have made clear initiatives to promote and develop renewable energy production. This is beginning to be introduced to the urban wind turbine industry, where SWTs have been recognized as effective energy production systems for reducing GHG emissions. Although there are still some technical and financial challenges to overcome to achieve larger utilization of SWTs, these should be able to be solve d at least in part in the coming years, with more research on modelling urban environments and specific built structures to optimize the placement of SWTs. Finally, further government-sponsored schemes will allow the development and implementation of SWTs into the urban environment to occur rapidly.

References

[1] Polman A, Knight M, Garnett E C, Ehrler B and Sinke W C 2016 Photovoltaic materials: present efficiencies and future challenges *Science* **352** aad4424

[2] ISE Fraunhofer Institute for Solar Energy Systems 2017 Photovoltaics report [Internet] [cited 2017 Dec 6]. Available from: https://ise.fraunhofer.de/content/dam/ise/de/documents/publications/studies/Photovoltaics-Report.pdf

[3] Green M A *et al* 2017 Solar cell efficiency tables (version 50) *Prog. Photovoltaics Res. Appl. [Internet]* **25** 668–76

[4] Solar Media Finlay Colville 2017 Thin-film solar production to collapse to seven-year low in 2017

[5] Fu R, Feldman D, Margolis R, Woodhouse M and Ardani K 2017 US solar photovoltaic system cost benchmark: Q1 2017 *EERE Publication and Product Library*

[6] Celik I *et al* 2017 Energy payback time (EPBT) and energy return on energy invested (eroi) of perovskite tandem photovoltaic solar cells *IEEE J. Photovoltaics* **8** 305–9

[7] Trend Force Corporation 2017 PV Energy Trends

[8] First Solar First Solar Series 4 PV Module [Internet]. 2017 [cited 2018 Dec 11]. Available from: http://firstsolar.com/Modules/Series-4

[9] Sendy A 2017 Pros and cons of string inverters vs micro inverters Solar Reviews

[10] Tamboli A C, Bobela D C, Kanevce A, Remo T, Alberi K and Woodhouse M 2017 Low-cost CdTe/silicon tandem solar cells *IEEE J. Photovoltaics* **7** 1767–72

[11] REN21 2016 Renewables 2016: global status report; REN21 Secretariat Paris

[12] Musial W and Ram B 2010 Large scale on-shore wind power in the United States: assessment of opportunities and barriers NREL/TP-500-40745. Available from: https://www.nrel.gov/docs/fy10osti/40745.pdf

[13] Mertens S 2006 Wind energy in the built environment: concentrator effects of buildings (Brentwood, UK: Multi-science)

[14] Balduzzi F, Bianchini A, Carnevale E A, Ferrari L and Magnani S 2012 Feasibility analysis of a Darrieus vertical-axis wind turbine installation in the rooftop of a building *Appl. Energy [Internet]* **97** 921–9

[15] Krishnan A and Paraschivoiu M 2016 3D analysis of building mounted VAWT with diffuser shaped shroud *Sustain Cities Soc.* **27** 160–6

[16] Islam M R, Mekhilef S and Saidur R 2013 Progress and recent trends of wind energy technology *Renew. Sustain Energy Rev.* **21** 456–68

[17] Dominy R, Lunt P, Bickerdyke A and Dominy J 2007 Self-starting capability of a Darrieus turbine *Proc. Inst. Mech. Eng. Part A J. Power Energy* **221** 111–20

[18] Predescu M 2016 Economic evaluation of small wind turbines and hybrid systems for residential use *Renew. Energy Environ. Sustain* **1** 33

Chapter 6

Energy storage

Jacob J Lamb, Bruno G Pollet and Odne S Burheim

Energy storage is essential for an energy system to function with high efficiency and flexibility. Engineering a system to store energy to fulfil these requirements is paramount. If intermittent renewable energy sources, are being utilized, it is important that the electrical energy produced is stored to allow distribution when needed irrespective of the light and wind conditions. There are several technologies that have been engineered for storing energy, and each of them has their own strengths and weaknesses for different applications. In this chapter, a select few technologies are presented outlining their technology and what energy they store.

6.1 Biomass

During the twentieth century, fossil sources turned into the primary energy supply for transportation. This persisted with most transport vehicles operating on fossil-fuel derivatives into the twenty-first century [1]. Continued use of these energy sources to meet the worldwide requirement for energy is problematic as it raises atmospheric CO_2 levels, causing climate change [2]. In addition, fossil fuels are a finite resource that is limited in fixed locations [3]. There has been an increasing interest in biofuels for some decades now, taking into consideration environmental, financial and political factors. Biofuels are produced from natural biomass and can be used as fossil fuel replacements. In addition, their combustion results in reduced sulfuric and aromatic exhaust emissions compared to typical fuels while offering comparable fuel efficiency characteristics [4].

The word biofuel reflects liquid or gaseous materials mainly produced from biomass. Large-scale biofuel manufacturing provides humanity with an opportunity to restrain our reliance on fossil use. In addition, there is an increasing tendency in developed countries to use state-of-the-art techniques and economical bioenergy transformation procedures using a range of biofuels that have become economically viable with fossil fuels [5].

doi:10.1088/978-0-7503-3259-0ch6

6.1.1 Development of bioenergy

Biomass, mostly wood, is the earliest kind of energy people used. Biomass has historically been used by immediate combustion and this technique is still commonly used today. To date, timber and timber landfill, agricultural crops and their by-products, urban solid waste, animal waste, food processing waste, aquatic plants and algae are the main vital forms of biomass energy. Biomass energy can be used directly in the electrical, lumber and timber manufacturing plants and pulp and paper sectors. Much assessment of safe and environmentally friendly energy from biomass products is currently targeted towards transition from fossil fuels.

6.1.1.1 Components

Within the procedures for producing necessary materials and chemicals, the chemical composition and major organic components in agriculture can be used directly [6]. The three polymeric parts are the constituents of lignocellulose: cellulose, hemicellulose, and lignin [7]. The composition of lignocellulose is stiff and obstinate, resulting in a rather challenging biochemical transformation into usable sugars [8–13]. The main three components generally have elevated molecular weights and add a lot of mass [14].

Cellulose originates in a variety of types with a substantial portion from domestic and industrial waste [15]. Hemicellulose may be a combination of various polymerized monosaccharides such as aldohexoses, mannose, galactose, xylose, arabinose, 4-O-methyl glucuronic acid and residues of galacturonic acid [16]. The backbone chain in hardwood (xylan) consists of xylose units paired with β-(1,4)-glycosidic bonds and branched with 4-O-methyl glucuronic acid groups by α-(1,2)-glycosidic bonds [14]. Lignin is a synthesized aromatic compound from precursors of phenylpropanoid. The vital chemical components of polymer phenylpropane (mainly syringyl, guaiacyl and p-hydroxy phenol) are held together by a set of matrix connections. This matrix includes a range of helpful components, such as methoxyl and carbonyl, which give the polymer organic compound a strong polarity [17].

6.1.1.2 Biomass importance

Due to the ongoing depletion of fossil fuels, bioenergy usage has attained clear interest in the past couple of decades [18]. Growing interest in bioenergy is motivated by [19]: (i) helping with poverty reduction in emerging nations; (ii) addressing power requirements in a cost-effective way; (iii) providing energy in a multitude of forms demanded by society (e.g. liquid and gaseous fuels for thermal and electricity supplies); (iv) small CO_2 footprint; and (v) helping to restore non-productive lands. As worldwide energy policies and strategies concentrate strongly on decarbonization, the significance of biomass in the energy transition from fossil fuels will increase.

6.1.1.3 Contribution of bioenergy to the global energy supply

Bioenergy is regarded as an optimal option as a supply of transitional energy between reducing fossil reliance and adapting a community based on hydrogen. Most significantly, for the near future, it can fulfil societies' energy demands. Biomass energy supplies are regarded to be the largest source of energy in the world. It is a natural resource comprising of about 220 billion dry tons (about 4500 EJ) of annual primary production [20], yet its utilization as an energy source for our society is currently limited. While access to and use of biomass is not straightforward, the long-term capacity for sustainable power use is expected to achieve 150–400 EJ/yr by 2050 with the use of woods and agricultural waste, as well as possibly increasing permanent energy crops [21].

The present transformation paths of biomass to fuels and electricity are characterized in heat, chemical and aromatic synthesis transformation paths [22], These include technologies for thermo-chemical conversion such as pyrolysis [23], gasification [24], liquefaction [25] and supercritical gasification of water [26] to maximize yields. Biomass thermo-chemical transformation systems such as pyrolysis and gasification are by no means the most important choices currently, as biomass combustion accounts for over 90% of worldwide bioenergy manufacturing [27]. The most used biomass methods in industrialized countries are the immediate combustion of residues and waste for generation of electricity [28]. Currently, biogas is the favoured biomass utilization (excluding biomass combustion) in Nordic countries.

6.1.2 Biogas fuel storage

Four types of pressure vessels can be used for storing pressurized gases [29].

1. Fully metallic vessels (up to 50 MPa).
2. Steel vessel with a glass fibre composite overwrap (high pressure).
3. Fully composite wrap with metal liner (reliable for 45 MPa working pressures).
4. Fully composite (up to 100 MPa).

In a perfect world, biogas would be upgraded and directly introduced into the infrastructure of natural gas. It becomes apparent that liquid storage is more favourable for high-quality biogas when given the necessary improvements in purity after anaerobic digestion. One of the main parts of gas liquification systems are the heat exchangers. Cryogenic heat exchangers should be intended to work at the smallest feasible temperature variations in order to obtain optimal overall energy efficiency of the system. This needs big heat transfer areas and large, high-cost equipment [30, 31]. The heat transfer occurs in many parts of the process [32], which can be summarised as [33]:

1. process cooling;
2. separation of gas mixtures; and
3. liquefaction for transportation and storage.

Liquefaction of biogas reduces gas volume by nearly 600 times, making storage and transportation easier and safer. The technology of liquified biogas (LBG) offers an economically viable manner to transport of biogas vast distances. Biogas liquefaction requires large exchangers of cryogenic heat, where biogas is cooled, liquefied and subcooled to 111 K. Some liquefaction systems use distinct heat exchangers to fulfil the obligation of precooling or subcooling, while others may involve only one multi-stream heat exchanger per manufacturing cycle, frequently known as the primary heat exchanger [34]. In view of the related capital and operating costs, the economic feasibility of many LBG projects is largely governed by the efficiency of the primary cryogenic heat exchanger [34]. Once delivered, at re-gasification facilities, LBG can be transformed back into biogas, generally requiring large heat exchangers.

6.2 Hydrogen

The majority of energy-related experts think that H_2 will play a major part in the future as a very significant main energy storage and carrier [35, 36]. H_2 is pure fuel without direct CO_2 emissions that can be used immediately to generate electricity in fuel cells. In addition, H_2 involves a large energy output of 122 kJ g^{-1}, 2.75 times greater than fossil fuels [37]. As an environmentally friendly energy [38], the use of H_2 as a fuel for transportation and storage receives significantly favourable publicity and has been widely studied for use in combustion motors and hybrid fuel cell vehicles. However, H_2 is in the form of a gas at traditional temperatures and pressures, presenting larger transport and transport hurdles than those present for liquid fuels [39]. Nevertheless, innovations have shown that in many solid and liquid compounds (e.g. metal hydrides, methane, methanol and lightweight hydrocarbons) H_2 can be combined into other chemicals [40].

Hydrogen is already commonly used for several procedures in a number of applications such as the chemical sector (63%) and refineries (31%) [41]. The production and use of hydrogen usually occurs within the same industrial facility. Security, logistics and financial elements drive this decision [41]. Hydrogen has also been used in the aerospace industry over the past sixty years as both propellant [42] and fuel cell technologies to generate electrical energy [43]. Hydrogen is stored in liquid phase for this purpose. Both the production and consumption of hydrogen are anticipated to increase in the coming years [44] primarily due to the hydrogen storage capacity combined with renewable energies. In this instance, it is possible to reduce hydrogen expenses by enhancing the energy efficiency of the entire life cycle of hydrogen. In addition, the elements of hydrogen safety would become more significant as unusual accident situations could occur that have not yet been regarded for hydrogen.

6.2.1 Development of hydrogen

Hydrogen is produced through various manufacturing techniques from large primary energy sources. Currently most gas is produced from unrenewable fossil fuel. Half of the present manufacturing of H_2 is based on thermo-chemical techniques that exploit fossil fuel as a source of energy. Of this, the biggest energy availability for H_2 manufacturing is provided by petroleum and natural gas followed by coal [45, 46]. In brief, H_2 production based on fossil fuel makes up 95% of worldwide manufacturing.

H_2 production from fossil fuels causes greenhouse emissions to co-produce, which are recognized to cause global climate change. H_2 generated by a multitude of natural primary energy forms such as wind, biomass and solar energy are perfect for fossil fuels' step-by-step replacement [47]. Biomass and biomass-derived fuels can be used to sustainably produce H_2. In addition, the use of biomass to produce H_2 rather than fossil fuels decreases the worldwide amount of CO_2 emissions released into the atmosphere. This is because the CO_2 emissions released from biomass were once consumed and fixed from the atmosphere in the growing plants' own biochemical reactions [15].

The present global annual utilisation of H_2 fuel is around 400–500 billion Nm^3 [15]. The current use of H_2 has a growth level that can be calculated at 5%–10% per year [45]. However, only a portion of this H_2 is currently used for energy storage and transport purposes; the bulk is used as a chemical feedstock for organic compounds, food and other process industries. The global marketplace for H_2 is already greater than US$ 40 billion per year [48]; it consists of use in ammonia production (49%), fossil fuel processing (37%), wood production (8%) and various smaller uses (6%) [49].

An advantage of H_2 is that it can be used as a transport fuel, whereas there is no direct use of either nuclear or renewable energy. H_2 is great for power storage and fuel transport when combined with fuel cells, since H_2 can be transformed directly into electricity anywhere, anytime.

6.2.2 Hydrogen storage

H_2's low density as a gas is one of the primary drawbacks. Compared to other gases, if placed under ambient circumstances, H_2 needs a larger volume. The quantity of space is a critical problem in most applications. Different storage methods of H_2 have been researched in recent decades for this purpose. To boost its concentration, H_2 can be compressed or liquefied. Other storage methods envisage binding it through physical adsorption with porous material such as metal–organic frameworks (MOF) or zeolites. In addition, H_2 can be placed in metal hydrides or liquid organic H_2 carriers (LOHCs) via a chemical bond (absorption). An evaluation of H_2 storage can be found in the literature [50]. In the following, only the compressed and liquid H_2 solutions are described.

6.2.2.1 Compressed gaseous hydrogen (CGH_2)

The most useful and popular method of hydrogen storage is compressed hydrogen. The compression can boost the density of hydrogen at 700 bar and room temperature [51] up to 39.2 kg m^{-3}. As the hydrogen pressure rises, the compression energy demand also rises. In theory, 1.05 kWh/kgH_2 is required to compress hydrogen from 20 to 350 bar isothermally, whereas 1.36 kWh/kgH_2 is required to compress it up to 700 bar [52]. Therefore, the compression energy required is not linear. In theory, the compressor consumes 1.7 to 6.4 kWh/kgH_2. In addition, the original cost of the machinery (e.g. compressors, tanks, tubes and valves) rises with increasing desired pressure.

There are currently four kinds of tanks for hydrogen (Type I, II, III and IV) [29]. Usually metal, composite (carbon fibre and resin) and plastic are used in these tanks. The various kinds of containers can withstand varying stresses. When using H_2 to decrease expenses and meet safety requirements, the selection of components should be precise.

6.2.2.2 Liquid hydrogen (LH_2)

LH_2 is an accepted alternative for large quantities of hydrogen. It has a density of 70.9 kg m^{-3} at 20.4 K [51]. H_2 can easily evaporate due to its low boiling temperature (−253 °C). For this reason, the LH_2 containers consist of two metal tanks separated by a vacuum jacket packed with insulation fabric [53]. The heat losses with the atmosphere are significant enough to cause the LH_2 to evaporate, even if the LH_2 tank is very well insulated. Therefore, the formation of boil-off gas (BOG) is one of the primary disadvantages of the storage technique of LH_2.

Hydrogen is comprised of 75% ortho-and 25% para-hydrogen under atmospheric conditions. Usually, hydrogen is transformed to 100% para-hydrogen during liquefaction to decrease the BOG level. This conversion is performed by a catalyst and is an exothermic process, therefore requiring removal of heat. The theoretical energy for ambient liquefaction is 3.3 kWh/kgH_2 with, and 3.9 kWh/kgH_2 without the conversion from ortho- to para-LH_2.

In practice, the liquefaction energy requirements are 10–13 kWh/kgH_2 [52]. This quantity of electricity is equivalent to 30%–40% of the lower heating value (LHV) of hydrogen, which is 120 MJ/kg (33.3 kWh kg^{-1}). A portion of the heat could be retrieved during the LH_2 evaporation; therefore, the total energy consumption could be reduced. In this situation, the material selection should take into account the low temperature effects as well. Despite all these disadvantages, due to the elevated energy density relative to other hydrogen storage solutions, interest in this form of storage is growing rapidly.

6.3 Present uses of hydrogen and biomass

6.3.1 Heat and industry

Heating and lighting represent approximately 60%–80% of power consumption in housing and commercial structures across Europe. To avoid an environmental catastrophe, greenhouse emissions must be decreased significantly swiftly and mostly eliminated by 2050. However, for a variety of reasons the electricity industry in many areas (e.g. the UK) is extremely hard to decarbonize [54]:

1. Heating is the largest energy demand;
2. The requirements for heating vary significantly due to location, season and time of day;
3. Heating from fossil fuels provides flexibility.

A specific challenge in temperate countries for low-emission heating is fulfilling the high requirement for winter heating, which is considerably greater than the peak demand for non-heating electricity. Traditional gas heating systems meet the large seasonal variation between summer and winter, as their small cost per kW and convenience of storage and transportation render them perfect. However, their CO_2 emissions are a major adverse result of their use. Improved insulation, domestic or community thermal storage and energy storage systems can help scale back the peak demands to decrease the usage of these energy forms; however, in several nations, the robust strategy needed to advocate these adaptations has not yet been implemented. Moreover, although multiple low-carbon options such as electrification or district heating may satisfy peak heating requirements, it would require enormous investment in infrastructure.

Natural gas is widely used worldwide and is currently a cheap, easy and relatively clean source of energy relative to coal and oil. Because of this, in several counties, it is the dominant heating fuel. Biogas is also used and directly introduced into natural gas networks in several areas to deliver heating energy. In addition, district heating is widely used in Russia and Scandinavia [55], allowing for concentrated use of energy, which is a simple solution for the distribution of biogas.

6.3.2 Infrastructure

Developing the infrastructure for H_2 and biofuel is a very significant obstacle to allow the extensive use of these energy sources. There is a perception that, without using current energy infrastructure, a 'hydrogen economy' would be developed at a colossal price. However, there are already diverse manufacturing and delivery paths (e.g. reforming and methanation) and many progressive measures that do not require complete infrastructure changes (e.g. the use of pre-existing natural gas equipment). By using the present infrastructure, developing a cost-effective infrastructure that evolves from these resources will enable a fully decarbonized energy sector to be established that our society can transition to over time.

6.4 Heat energy storage

Storing heat can be convenient for several purposes. The motivation factor for heat storage is to lower peak power demands or to save energy. For example, during day cycles, having access to heat during peak hours when the electrical energy is otherwise expensive is ideal. Having access to cheap heat in the winter and cheap cooling in the summer is an example of seasonal heat storage. District heating and water tanks are common ways to store heat for daily fluctuations. In the case of distributed heating systems, waste heat from nearby industry and waste incinerators can be used to heat water that is subsequently pumped into a distributed network of pipelines. The temperature in such pipelines can change over time. When consumers wake up and need warm water for showering and other commodities, the temperature of the district heat network lowers. This is because more heat is drawn than added.

6.4.1 Electrical hot water heaters

Domestic hot water (DHW) heaters use electricity to heat water and retain energy for the purpose of use within the building. The hot water tank based on the heating of water is one of the most common energy storage systems. They comprise an isolated storage tank where hot water is stored. A thermal unit generates warm water outside or inside the isolated tank. The energy saved relies on the temperature of the warm water and the size of the tank, and the insulation of the tank determines the thermal loss and limits the period of storage.

A standard hot water heater works according to a thermostat without intelligent command. It will always try to keep the set point temperature, usually around 80 °C. However, usually water temperatures around 50 °C are the highest temperatures required for warm water purposes in buildings. Hot water is generated between 55 °C and 65 °C for utilisation in space heating. The temperature for sanitary hot water heating usually ranges from 60 °C–70 °C to prevent development of unwanted microorganisms.

The hot water heater should never reach below 55 °C when considering the heat losses in the storage scheme. For the dimensioning of DHW systems, the cold city water temperature is typically around 10 °C in Nordic countries, and the minimum temperature in the hot water tank is 55 °C to satisfy the necessary requirements of a standard domestic building [56]. This hot water storage can be considered empty if the hot water heater reaches less than 55 °C.

Storage of hot water energy is a mature technology used in Europe and around the world on a large scale. In both housing and tertiary sectors, renewable energy systems such as solar thermal and heat pumps are increasingly being used to generate the DHW. The tank-in-tank technology has been created in recent years for devices offering space heating and DHW in residential buildings. In this case, the DHW is contained in a small tank within a large tank which stores thermal energy for heating the smaller tank. Using one boiler only allows the system to be more affordable.

The storage tank efficiency and the heat losses are extremely dependent on the tank isolation and the flow rate of water. The flow rate relies on the indoor temperature when the warm water is used for room heating. When using the heated water for sanitary water heating, the flow rate depends on the amount of water per time unit being used.

Although the individual thermal storage capacity is low, thermal storage is installed in very high numbers. This means that the general storage capability across a large area (region or country) could be very significant and can be used with great ease to capture maximum electricity output from renewable sources such as wind or photovoltaics.

6.5 Energy storage by batteries

Batteries are devices that contain chemicals suited to storage of electric energy. We define batteries as either primary or secondary [57]. Primary batteries are disposable and are used once and then dispatched. Secondary batteries are rechargeable and can store electrical energy and release this energy on demand.

In planning energy storage, there are two topics that are of importance: state of charge (SoC) and state of health (SoH). SoC refers to the percentage of the battery capacity that is available, not unlike fuel level in a fuel tank. SoC is the percentage level typically indicated in the upper part of a cellular phone or computer screen. Sometimes, also the term depth of discharge (DoD), is used instead of SoC. DoD in percentage is 100 − SoC in percentage so that DoD and SoC always sum up to hundred percent. SoH refers to the amount of the initial capacity that is available. In the fuel level analogy, the tank shrinks with time. Batteries don't shrink, but some of the reactants becomes immobile as the battery is charged and recharged several times.

An important term for secondary battery engineering is the C-rate. The C-rate is defined as how fast a battery is charged relative to a full equivalent circuit in one hour. Moreover, charging a battery from 40%–90% in one hour is half a C (C/2), and charging a battery from 0%–100% in half an hour is 2C. Fast charging rates for a car are typically much faster than 1C, while charging slower than C/2 is considered very slow.

Several types of secondary batteries are available for energy storage and they each have different characteristics. The differences easiest to recognize between secondary batteries are the specific density (e) and energy density ($\hat{e}$). Table 6.1 gives an overview of batteries relevant for storing energy in buildings. Obviously other battery technologies are available too, and Li-ion is the most modern and reasonable choice; however, lead acid batteries and nickel metal hydride (NiMeH) are of contextual interest too [57].

The lead acid battery is a well-established technology that has been commercially available for almost a century. Its main market is for supplying power to starting engines and supporting the voltage and frequency of electricity in uninterrupted power supply (UPS) smart grids. The lead acid battery has become cheap and can be used for energy storage in off-grid units where little energy is needed (e.g. lights, TV, radios and TVs in boats and cabins). When there is a need for larger amounts of energy and not just a short-term power supply, then batteries with higher energy density are required. This is true at least for the transport sector where space and weight are a definite limitation. In buildings, space and weight is important too. Space has a value, so when energy density can be ten times smaller for lead acid

Table 6.1. Battery technology and property overview [58–60].

Battery type	Specific energy (kWh kg^{-1})	Specific power (kWh kg^{-1})	Price (\$ kWh^{-1})	Energy density (Wh L^{-1})
Lead	20–40	20–200	100–150	50–100
NiMeH	50–70	10–100		140–300
Li-ion	150–200	100–500	150–250	250–550

batteries compared to Li-ion ones, this could be an important factor when choosing the battery technology. Inconvenient space like attics and basement corners can be used for computer servers and energy storage rooms, but sometimes only to a limited degree. Beyond this, space that can otherwise be rented out is accumulated and then the cost of space is added on top of the batteries.

The nickel metal hydride battery (NiMeH) supplies much more energy than the lead acid one. It is based on advancing the nickel cadmium technology by using the hydrogen reaction and storing hydrogen in a metal hydride. The technology has been used for many applications and is well understood. Furthermore, several hybrid electric vehicles use NiMeH batteries for this reason. Despite its energy density, it can be attractive for storing at least moderate amounts of energy. The battery requires full cycling due to its memory effect, where, if not fully cycled, parts of the capacity will be temporarily lost.

The lithium ion battery (LiB) is based on lithium mixed with transition metal oxides on one electrode and metallic lithium stored in carbon on the other. When fully charged, the lithium mixed metal oxide electrode is depleted of lithium and the carbon anode is filled up with metallic lithium. During discharge, the transition metals in the cathode are reduced while lithium in the carbon is oxidized into lithium ions, which simultaneously migrate to the cathode via an organic electrolyte separating the two electrodes. LiB is an established technology matured to the level where most electric (fully, plug-in and hybrid) vehicles use it as a means for the storage of energy needed for propulsion. Even ferries using LiB in commercial traffic are established as fully electric vehicles. Therefore, the technology is ready for being used in buildings for storing energy. When considering LiB, one must be aware that a large family of batteries exists and the main difference between these are the transition metal oxides mixed with the lithium ions in the cathode. The first commercial LiB was based on cobalt oxide ($LiCoO_2$) and is the most energy intensive and expensive LiB. The cheapest and least energy intensive LiB to produce is based on iron phosphate ($LiFePO_4$). More about the different types of batteries can be found in the literature [57].

A topic to consider when choosing the battery type in addition to energy density and specific energy is the battery lifetime. For lead acid batteries the expected lifetime is often in the range of 3–5 years and for Li-ion 3–15 years. After such time the SoH of the battery is lowered to a level where too little energy is stored for the installation to make sense. The SoC window selected affects the ageing of any battery, in particular with the lead acid battery and LiBs. As a simple example, consider a lead acid battery to have a life of 3 years when using an SoC window 0%–100% and 5 years when cycled between SoC of 20%–90%. Considering a representative price of 125 USD per kWh, the energy storage cost (at 100% energy efficiency) for the 3 and 5 year case becomes 0.114 and 0.98 USD kWh, respectively. As much as one saves money in the long-term installation, this only makes sense if one has 43% extra battery capacity and otherwise unusable space for storage of the batteries. The short-term example considers 1095 full cycles and the second one 1278 normalized cycles. It is common for lead acid and other battery technologies, that limiting the cycle SoC window leads to increased amounts of normalized cycles.

Cycle normalization is really as simple as the part of the utilized SoC window, where for the example, two cycles between 30% and 80% SoC would equal one normalized cycle. As much as this example is representative for batteries for energy storage, it should not be taken literally, and the number of expected normalised cycles from different SoC window usage changes with specific batteries, temperature and specification information for each battery product.

An important aspect when having batteries for energy storage is temperature control. The most important factor that reduces battery lifetime is temperature increases. As a guideline for all organic based electrolytic cells (and most chemical degradation processes) is that lifetime is lowered by a factor of two for every 10 °C temperature increment. This is as the ageing rate mechanisms follow an Arrhenius behaviour at least to a first order approximation [61]. Therefore, battery cooling is required; however, this can be good news in the cases where heat is needed elsewhere in the building. For instance, air can be passed through a battery storage room on the way into an air distribution system.

A third aspect that affects ageing is charging and discharging rates, and typically one must go above C-rates of 1C to observe significant effects of charging and discharging rates [61]. The discharge rate is more detrimental to the battery SoH than the charging rates; however, both are less detrimental than an increased temperature. When charging and discharging rates are significantly lower than 1C, the ageing mechanisms are considered the same as if the battery is left on a shelf by itself, and the temperature and SoC appear to be the only important factors for the reduction in SoH.

One of the greatest things about batteries and LiBs in particular is their energy efficiency. When charging or discharging rates are significantly slower than 1C, the efficiency of a LiB is up to 98%. Charging and discharging at rates of C/2 or less, it is reasonable to expect a cycle efficiency of 96% or better. In this light, it makes sense to also purchase electric energy at periods of the day when it is cheap, as long as the difference between the peak and lowest price is significant enough.

References

[1] Loppacher L J and Kerr W A 2005 Can biofuels become a global industry?: government policies and trade constraints *Energy Polit.* **5** 7–27

[2] Balat M 2009 Bioethanol as a vehicular fuel: a critical review *Energy Sources, Part A* **31** 1242–55

[3] Demirbas M F 2009 Bio-oils from corn stover via supercritical water liquefaction *Energy Educ. Sci. Technol. Part A-Energy Sci. Res.* **23** 97–104

[4] Demirbas A and Dincer K 2008 Sustainable green diesel: a futuristic view *Energy Sources, Part A* **30** 1233–41

[5] Demirbas A 2008 The importance of bioethanol and biodiesel from biomass *Energy Sources, Part B* **3** 177–85

[6] Balat M 2007 Hydrogen in fueled systems and the significance of hydrogen in vehicular transportation *Energy Sources, Part B Econ. Planning, Policy* **2** 49–61

[7] Yaman S 2004 Pyrolysis of biomass to produce fuels and chemical feedstocks *Energy Convers Manag.* **45** 651–71

[8] Lamb J J, Islam M H, Hjelme D R, Pollet B G and Lien K M 2019 Effect of power ultrasound and Fenton reagents on the biomethane potential from steam-exploded birchwood *Ultrason. Sonochem. [Internet]* 104675 Available from: http://sciencedirect.com/science/article/pii/S1350417719308909

[9] Horn S J, Vaaje-Kolstad G, Westereng B and Eijsink V 2012 Novel enzymes for the degradation of cellulose *Biotechnol. Biofuels* **5** 45

[10] Quinlan R J *et al* 2011 Insights into the oxidative degradation of cellulose by a copper metalloenzyme that exploits biomass components *Proc. Natl. Acad. Sci.* **108** 15079–84

[11] Vaaje-Kolstad G *et al* 2010 An oxidative enzyme boosting the enzymatic conversion of recalcitrant polysaccharides *Science* **330** 219–22

[12] Vivekanand V, Olsen E F, Eijsink V G H and Horn S J 2013 Effect of different steam explosion conditions on methane potential and enzymatic saccharification of birch *Bioresour. Technol.* **127** 343–9

[13] Sarker S, Lamb J J, Hjelme D R and Lien K M 2019 A review of the role of critical parameters in the design and operation of biogas production plants *Appl. Sci.* **9** 1915

[14] Demirbas A 2009 Pyrolysis of biomass for fuels and chemicals *Energy Sources, Part A* **31** 1028–37

[15] Balat H and Kırtay E 2010 Hydrogen from biomass–present scenario and future prospects *Int. J. Hydrogen Energy* **35** 7416–26

[16] Mohan D, Pittman C U and Steele P H 2006 Pyrolysis of wood/biomass for bio-oil: a critical review *Energy & fuels* **20** 848–89

[17] Feldman D, Banu D, Natansohn A and Wang J 1991 Structure–properties relations of thermally cured epoxy–lignin polyblends *J. Appl. Polym. Sci.* **42** 1537–50

[18] Caputo A C, Palumbo M, Pelagagge P M and Scacchia F 2005 Economics of biomass energy utilization in combustion and gasification plants: effects of logistic variables *Biomass Bioenergy* **28** 35–51

[19] Karekezi S, Lata K and Coelho S T 2006 Traditional biomass energy: improving its use and moving to modern energy use *Renew. Energy—A Glob. Rev. Technol. Policies. Mark* **1** 231–61

[20] Jekayinfa S O and Scholz V 2009 Potential availability of energetically usable crop residues in Nigeria *Energy Sources, Part A Recover Util. Environ. Eff.* **31** 687–97

[21] Charles M B, Ryan R, Oloruntoba R, von der Heidt T and Ryan N 2009 The EU–Africa energy partnership: towards a mutually beneficial renewable transport energy alliance? *Energy Policy* **37** 5546–56

[22] Hoogwijk M, Faaij A, Eickhout B, de Vries B and Turkenburg W 2005 Potential of biomass energy out to 2100, for four IPCC SRES land-use scenarios *Biomass Bioenergy* **29** 225–57

[23] Caglar A 2009 Valorization of tea wastes by pyrolysis *Energy Educ. Sci. Technol. Part A-Energy Sci. Res.* **23** 135–44

[24] Bhuiyan M N A, Ota M, Murakami K and Yoshida H 2009 Pyrolysis kinetics of newspaper and its gasification *Energy Sources, Part A Recover Util. Environ. Eff.* **32** 108–18

[25] Demirbas A 2008 Liquefaction of biomass using glycerol *Energy Sources, Part A* **30** 1120–6

[26] Şener Ü, Genel Y, Saka C, Kiliçel F and Küçük M M 2009 Supercritical fluid extraction of cotton stalks *Energy Sources, Part A Recover Util. Environ. Eff.* **32** 20–5

[27] Demirbas A 2009 Biofuels from agricultural biomass *Energy Sources, Part A* **31** 1573–82

[28] Bakis R 2007 Alternative electricity generation opportunities *Energy Sources, Part A Recover Util. Environ. Eff.* **30** 141–8

[29] Barthelemy H, Weber M and Barbier F 2017 Hydrogen storage: recent improvements and industrial perspectives *Int. J. Hydrogen Energy* **42** 7254–62
[30] Pacio J C 2012 Multiscale thermo-hydraulic modeling of cryogenic heat exchangers *PhD Thesis* Norwegian University of Science and Technology
[31] Shah R K and Sekulic D P 2003 *Fundamentals of Heat Exchanger Design* (New York: Wiley)
[32] Barron R F 1999 *Cryogenic Heat Transfer* (Boca Raton, FL: CRC press)
[33] Fikiin K *et al* 2017 Refrigerated warehouses as intelligent hubs to integrate renewable energy in industrial food refrigeration and to enhance power grid sustainability *Trends Food Sci. Technol.* **60** 96–103
[34] Popov D *et al* 2019 Cryogenic heat exchangers for process cooling and renewable energy storage: a review *Appl. Therm. Eng.* **153** 275–90
[35] Veziroğlu T N and Şahi S 2008 21st Century's energy: hydrogen energy system *Energy Convers Manag.* **49** 1820–31
[36] Balat M 2008 Potential importance of hydrogen as a future solution to environmental and transportation problems *Int. J. Hydrogen Energy* **33** 4013–29
[37] Kapdan I K and Kargi F 2006 Bio-hydrogen production from waste materials *Enzyme Microb. Technol.* **38** 569–82
[38] Cherry R S 2004 A hydrogen utopia? *Int. J. Hydrogen Energy* **29** 125–9
[39] Ulusarslan D, Gemici Z and Teke I 2009 Currency of district cooling systems and alternative energy sources *Energy Educ. Sci. Technol. Part A-Energy Sci. Res.* **23** 31–53
[40] Shakya B D, Aye L and Musgrave P 2005 Technical feasibility and financial analysis of hybrid wind–photovoltaic system with hydrogen storage for Cooma *Int. J. Hydrogen Energy* **30** 9–20
[41] Ausfelder F and Bazzanella A 2016 Hydrogen in the chemical industry *Hydrogen Science and Engineering: Materials, Processes, Systems and Technology* (Hoboken, NJ: Wiley), 19–39
[42] Cecere D, Giacomazzi E and Ingenito A 2014 A review on hydrogen industrial aerospace applications *Int. J. Hydrogen Energy* **39** 10731–47
[43] Baroutaji A, Wilberforce T, Ramadan M and Olabi A G 2019 Comprehensive investigation on hydrogen and fuel cell technology in the aviation and aerospace sectors *Renew. Sustain. Energy Rev.* **106** 31–40
[44] International Energy Agency 2017 Global Trends and Outlook for Hydrogen
[45] Mohan S V, Bhaskar Y V and Sarma P N 2007 Biohydrogen production from chemical wastewater treatment in biofilm configured reactor operated in periodic discontinuous batch mode by selectively enriched anaerobic mixed consortia *Water Res.* **41** 2652–64
[46] Das D, Khanna N and Veziroğlu N T 2008 Recent developments in biological hydrogen production processes *Chem. Ind. Chem. Eng. Quarterly/CICEQ* **14** 57–67
[47] Yolcular S 2009 Hydrogen production for energy use in European Union countries and Turkey *Energy Sources, Part A* **31** 1329–37
[48] Kraus T 2007 Hydrogen Fuel—An Economically Viable Future for the Transportation Industry? *Duke J. Econ.* **19** 39
[49] Konieczny A, Mondal K, Wiltowski T and Dydo P 2008 Catalyst development for thermocatalytic decomposition of methane to hydrogen *Int. J. Hydrogen Energy* **33** 264–72
[50] Burheim O S 2017 Hydrogen for energy storage *Engineering Energy Storage* (New York: Academic), pp 147–92
[51] NIST 2019 NIST Chemistry WebBook

[52] DOE 2009 Energy requirements for hydrogen gas compression and liquefaction as related to vehicle storage needs

[53] Peschka W 1992 *Liquid Hydrogen – Fuel of the Future* 1st edn (Wien: Springer)

[54] Chaudry M, Abeysekera M, Hosseini S H R, Jenkins N and Wu J 2015 Uncertainties in decarbonising heat in the UK *Energy Policy* **87** 623–40

[55] Hanna R F, Gross R and Parrish B 2016 Best practice in heat decarbonisation policy: a review of the international experience of policies to promote the uptake of low-carbon heat supply

[56] Fan J 2009 System for Domestic Hot Water Services. Educational Notes for 11127 Sustainable Heating and Cooling of Buildings

[57] Burheim O S 2017 *Engineering Energy Storage* (New York: Academic)

[58] Goodenough J B and Park K-S 2013 The Li-ion rechargeable battery: a perspective *J. Am. Chem. Soc.* **135** 1167–76

[59] Wood D L III, Li J and Daniel C 2015 Prospects for reducing the processing cost of lithium ion batteries *J. Power Sources* **275** 234–42

[60] Gerssen-Gondelach S J and Faaij A P C 2012 Performance of batteries for electric vehicles on short and longer term *J. Power Sources* **212** 111–29

[61] Richter F, Kjelstrup S, Vie P J S and Burheim O S 2017 Thermal conductivity and internal temperature profiles of Li-ion secondary batteries *J. Power Sources* **359** 592–600

Chapter 7

Optimal control of batteries and hot water heaters in zero emission neighbourhoods

Martin Gjertsen, Jacob J Lamb, Inger Andresen and Kristian M Lien

This chapter presents a case-study on the energy flexibility of a zero emission neighbourhood. In order to assess the potential for reduction of peak loads and increasing the consumption of renewable energy produced on site, the optimisation of stationary batteries and electric hot water heaters has been modelled. This model was used to optimise for cost-savings in energy consumption and discusses the cost in relation to the CO_2 emissions of the energy consumed.

7.1 Introduction

The campus at Evenstad has been set up as a pilot zero emission neighbourhood (ZEN) project. The campus is a college campus located in Hedmark county in Norway. In order to achieve a ZEN status in terms of GHG emissions, campus Evenstad must account for the GHG emissions related to the buildings on campus through self-generation of renewable energy. An overview of the campus with its buildings can be seen in figure 7.1.

The Zero Emission Neighbourhood (ZEN) research centre is one of the research centres on environmentally benign energy (FME) in Norway. ZEN conducts research on Zero Emission Neighbourhoods in smart cities, and the goal is to develop solutions for future buildings and neighbourhoods with zero greenhouse gas (GHG) emissions. One of the key success factors to achieve this goal is identified by ZEN as being able to develop technologies and solutions for the design and operation of energy flexible neighbourhoods [1].

Today the campus has a 70 kWp PV-system, a 100 m^2 solar collector system, and a 40 kW_{el}/100 kW_{heat} wood chip fired combined heat and power plant (CHP), which provides renewable heat and electricity to the buildings at the campus. However, the campus is today far from accounting for all the energy related GHG emission, mostly due to the many old buildings and inefficient energy use on campus. This

doi:10.1088/978-0-7503-3259-0ch7

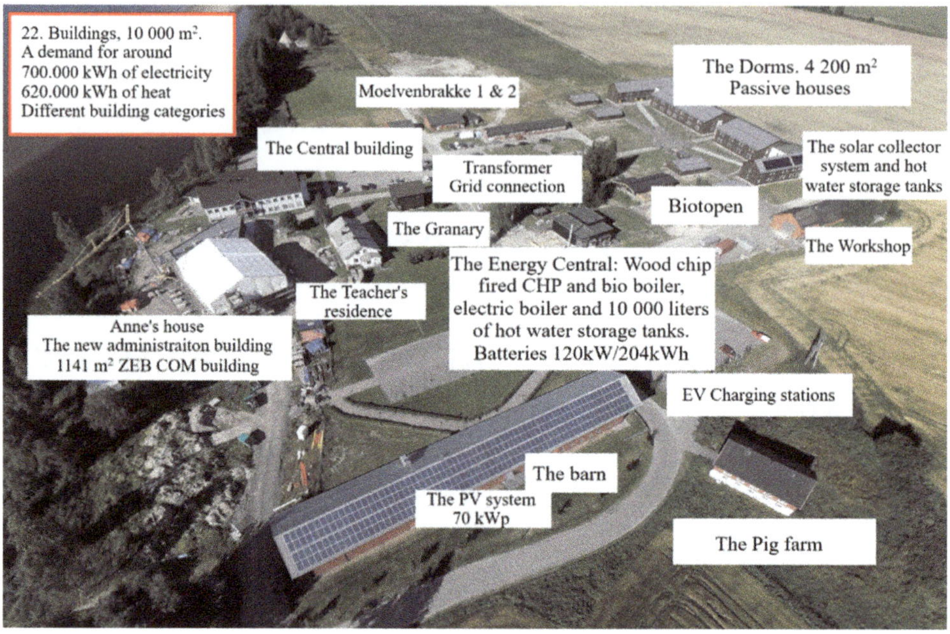

Figure 7.1. Overview of Campus Evenstad.

chapter will review the system at Evenstad as a case study to determine what is potentially the optimal method for control of the system.

7.1.1 Grid tariff structure in Norway

The current regulations give the grid companies a great degree of freedom in the design of tariffs. Most of the consumers today pay for the net consumption measured over a certain period (usually one to three months), regardless of the actual grid capacity the consumer demands. If the same regulation for the tariff structure is kept, the Norwegian Water Resource and Energy Directorate (NVE) expects that the power demand will continue to increase. As a result, the grid rental share of the electricity bill will increase by around 30% towards 2025. Most of the time, the grid will be heavily oversized. A more economical solution would be to reduce the peak loads in the system, reducing the increase in the grid tariff [2].

In November 2017, NVE published a consultation document, suggesting a change in the tariff structure of the distribution grid [2]. Two of the suggested tariffs were subscribed power and time of use. These tariffs should have the potential to reduce the peak demand in the electrical grid (the hours when the simultaneous consumption is largest). Reducing the peak loads will postpone or avoid grid investment and secure an economic operation of the grid. To avoid disturbance of the price incentive in the power market, NVE suggests that the tariff models should operate with the same time resolution.

The design of the grid tariffs was based on the weighted average energy tariff for households in Norway from 2017. First, the annual costs of the weighted average energy tariff were calculated for the example consumer. Secondly, the grid tariffs were designed so that the annual costs were the same as for the weighted average energy tariff. With a consumption of 22 300 kWh, the customer would currently pay 6075 NOK annually. The suggested new grid tariff models were then designed so that the annual costs would be close to 6075 NOK.

The grid tariff includes an energy part, an overconsumption part and a fixed part. In this grid tariff, the customer pays a fixed part depending on the size of the subscription. The size of the subscription yields the power threshold (PT). The PT determines the price of the electricity consumed. The customer pays only a small amount of money when electricity is consumed below the power threshold (BPT). If the consumer consumes electricity above the power threshold (APT), the consumer must also pay for the power consumed APT. When electricity is consumed APT it is referred to as overconsuming. When overconsuming, the consumer pays for the energy plus a price per kWh per h for each hour of overconsumption.

The energy part is set to the cost of the average marginal losses. The overconsumption part is set to 1 NOK per kWh per h. The fixed part consists of two parts, one for the power subscription size and one yearly fixed price. The subscription size is determined based on historical data for consumption. The power threshold is set so that the customer has 670 h of overconsumption. The price is set so that the fixed part and total costs are the same as the other tariffs. For the Example provided by NVE, the fixed part is 1060 NOK per year plus 689 NOK per kWh per h, which equals a total fixed cost of 1749 NOK per year. In the time of use grid tariff the grid companies increase the price for electricity in the energy part of the tariff, for the hours with expected peak hours. The electricity price in the energy part of the tariff is independent of the spot price.

7.1.2 Energy flexibility in buildings

The International Energy Agency (IEA) energy in buildings and communities (EBC) program carries out research toward near-zero energy and carbon emissions in the building environment. According to the ongoing project 'Energy Flexible Buildings', energy flexibility refers to a building's ability to change its demand and/or generation according to local climate conditions, user needs and grid requirements [3]. In process systems, flexibility represents the ability to accommodate process variations. Therefore, it is an important design objective.

Energy flexibility in buildings is according to some of the literature [3–5] assessed as the ability to change, or deviate from, the reference energy consumption profile. It can be achieved by either changing the energy consumption in time, or the quantity of power consumed. This presupposes that the consumer provides devices that can be intelligently controlled. Demand side management through intelligent devices can provide peak shaving and become very interesting for the distribution system operator (DSO).

It is well known that batteries are getting less expensive, and DSOs around the world are already investing in big stationary batteries due to flexibility issues. Bloomberg New Energy Finance (BNEF) predicts that the prices for lithium ion battery packs will fall by 52% between 2018 and 2030 [6]. In December they released a battery pack price survey, looking back on data from 2010. They could reveal that, between 2010 and 2018 the volume weighted average battery pack price had fallen 85%, from 1160 USD per kWh to 176 USD per kWh [7].

In contrast, IRENA investigated real prices for home storage battery systems offered by installers in Germany. In their report from 2017, IRENA stated a 60% median price decline from Q4 2014 to Q1 2017 [8]. Even though the total system cost was reduced from 2515 per kWh to 1017 per kWh, the latter price shows that commercialized residential battery systems still are quite expensive, and that the 176 USD per kWh suggested by BNEF does not apply for home storage systems [7].

7.1.3 Carbon emissions

In 2008, Norway set a target to reduce emissions by 12–14 million tons of CO_2 equivalents by 2020. In order to meet the Norwegian climate targets, Klima-og forurensningsdirektoratet (2010) investigated the potential and cost of possible measures for different sectors [9]. They found that a reduction of around 22 million tons of emissions can be achieved with measures at a cost below 4000 NOK per ton of CO_2 equivalents. To obtain a reduction of 12 million tons results in a marginal cost of around 1100 NOK per ton CO_2 equivalents. The figure shows that most of the measures in the building sector have a cost of around −500 to 2000 NOK per ton of reduced CO_2 equivalents.

7.1.4 Photovoltaic systems

When referring to the power output of the PV module, the unit kWp is commonly used. The p in kWp is 'peak' and refers to the output of the module under standard test conditions (STC). STC are composed of 1000 W per m^2 of irradiation at a 25 °C module temperature. STC is used when testing the PV modules for efficiencies. However, kWp can be a misleading unit since the solar panel yield is dependent on the incident light. A 100 kWp system in northern Europe can end up producing only half the annual energy yield compared to the same system installed in Chile. The unit kWh per kWp is often used to investigate the photovoltaic energy potential for an area.

The PV cells produce DC power; however, the electricity grid is most often an AC network. To convert the DC to AC an inverter is used. The inverter uses switches, switching at high speed to convert the DC to AC. In a PV system, the inverters will also match the voltage to the voltage of the network. If the inverter has one or more maximum power point trackers (MPPT), the inverter can find the optimal current–voltage characteristics of the PV system, obtaining the maximum power output.

The PV system can also be connected to a battery backup system. The batteries discharge DC electricity, so that the electricity can be used whenever there is a demand for it. The batteries can be connected directly to the PV system, without an inverter. However, in such a system the battery voltage will determine the voltage the PV system operates at. If the system is well designed, the system losses are small. With an inverter, or a charge controller with MPPTs in the system, the PV panels will operate at the optimum voltage, providing a more efficient PV and battery system.

7.1.5 Electric water heater

Electric hot water heaters for domestic hot water (DHW) use electricity to heat water. They consist of an insulated storage tank where the hot water is stored, and one or more electric heating elements. A regular hot water heater, without smart control, operates according to a thermostat. It will always try to maintain the set point temperature, most often around 80 °C. However, the highest required temperature for hot water purposes in buildings is usually around 50 °C.

In order to meet the desired demands for hot water when designing a DHW system, the minimum temperature in the hot water tank should be 55 °C [10]. If the hot water heater does not reach 55 °C, it can be considered as 'empty'. Typical tapping categories and suitable temperatures for bath, shower and hand wash are 40 °C, whereas kitchen wash is 45 °C [10]. The temperature difference between the water stored in the tank and the highest temperature required to meet the hot water purposes determines the energy quantity stored in the tank, calculated by:

$$Q_{DHW-tank} = V \cdot \rho \cdot cp \cdot \Delta T \tag{7.1}$$

$Q_{DHW-tank}$ is the energy stored in the tank, V is the tank volume, ρ is the density of water, cp is the specific heat capacity of water and ΔT is the temperature difference between the tank water temperature and the highest temperature demand for hot water purposes. When hot water is stored in DHW heaters, there will be heat losses due to the temperature difference between the water and the environment. These losses can be explained and quantified by:

$$Q_{loss} = U \cdot A \cdot \Delta T \tag{7.2}$$

where U is the heat transfer coefficient in W m^{-2} K^{-1} and A is the hot water heater surface area. The heat loss rate is proportional to the heat transfer coefficient and the ΔT. In order to minimize the heat loss, the tank is insulated, reducing the heat transfer coefficient [11]. Inside the hot water heater is one or several heating elements with a corresponding thermostat. The heating element converts electricity into heat through the resistive heating principle. The current passes through the heating element resulting in a change in electric potential and heat is dissipated to the water inside the tank [12]. When the hot water is drawn from the top of the tank, cold city water (or preheated water) immediately replaces the tapped water in the bottom of the tank. As the temperature of the water decreases and reaches a certain threshold temperature, the thermostat will switch on the heating element. The heating element is turned off as soon as the water reaches the desired temperature.

7.2 Case study of the campus at Evenstad

The energy flexibility in a Zero Emission Neighbourhood (ZEN) has been investigated. Campus Evenstad, which is a pilot project in the ZEN research centre has been used as the case study. The potential to reduce peak loads and increase self-consumption have been investigated through smart control of a 120 kW/ 204 kWh stationary battery and electric hot water heaters using an optimization model. The model was developed in Python. The neighbourhood was given a 433 kWp solar PV system simulated in PVSyst and the electricity consumption profiles were generated with the simulation software SIMIEN. The hot water consumption profiles were taken from the standard SN/TS 3031. The grid tariffs suggested by NVE, time of use and subscribed power, have been used as input to the model to see which of the grid tariffs works best in reducing the peak loads and increasing the self-consumption, with the objective of minimizing the electricity bill. In a third scenario, the CO_2 intensity in the electricity grid has been used as an input. In this scenario the model minimizes CO_2 emissions related to the electricity consumption to see how much the battery and the hot water heaters can reduce the carbon footprint in the ZEN. The three scenarios are compared to a baseline scenario with no flexibility.

7.2.1 The baseline scenario

In table 7.1, the electricity demand in the baseline scenario, production from the PV system and the resulting import and export of electricity is presented.

7.2.2 Electric water heaters

After visiting the campus, data from the hot water heaters in the eleven buildings were collected. The information as well as the results from the calculation of the useful energy and useful power can be seen in table 7.2. What is interesting to see from table 7.2, is that the heating elements in the hot water heaters on Campus Evenstad together add up to almost 100 kW. If all the hot water heaters were activated simultaneously for one whole hour, they would draw almost 100 kW of power from the grid. More importantly, if smart control is implemented, the electric hot water heaters can potentially reduce the power drawn from the grid by around 100 kW. With the current power purchase agreement, this would reduce the electricity bill by around 45 000 NOK annually for the neighbourhood. Even though the scenario of all the heating elements being activated simultaneously for one whole hour is unlikely, it shows that the potential for peak shaving is substantial, and that smart operation of the hot water heaters can be beneficial for both the consumer and the DSO.

Table 7.1. The electricity specifications for the baseline scenario.

Electricity demand	848 500 kWh
Electricity production	291 100 kWh
Self-generation	34.3%
Import	635 000 kWh
Export	77 500 kWh
Peak power demand	353 kW
Self-consumption	25.2%

Table 7.2. Electric water heater specifications.

Building	Tank volume [l]	Heating element	Useful energy stored [$a \cdot V$]	Useful power [$E \cdot \tau$]	Heat loss [75 °C–20 °C]
Anne's house	198	2 kW	12.79 kWh	1.80 kW	66 W
The barn	550	3 × 5 kW	35.52 kWh	13.50 kW	120 W
Moelvenbrakke 1	287	3 kW	18.54 kWh	2.70 kW	79 W
The central building	750	14 kW	48.44 kWh	11.40 kW	220 W
The teacher's residence	194	2 kW	12.53 kWh	1.80 kW	66 W
The welder's residence	200	2 kW	12.92 kWh	1.80 kW	100 W
The Workshop	100	1.95 kW	6.46 kWh	1.76 kW	70 W
Dorm 1 North	861	9 kW	55.61 kWh	8.1 kW	258 W
Dorm 1 South	1435	15 kW	92.68 kWh	13.5 kW	430 W
Dorm 2 North	1435	15 kW	92.68 kWh	13.5 kW	430 W
Dorm 2 South	861	9 kW	55.61 kWh	8.1 kW	258 W
Total	7071	89.95 kW	456.70 kWh	79.76 kW	2197 W

The neighbourhood has around 457 kWh of heat storage in the water heaters available for energy flexibility. This can be realised by postponing the power input, causing the temperature to drop from 90 °C towards 55 °C due to heat losses and hot water demand. The heat losses at 75 °C water temperature result in approximately 2.2 kWh of energy loss per hour from the hot water heaters on campus. However, the heat losses are only 0.5% when comparing to the total daily hot water demand of 457 kWh.

In figure 7.2, the hot water tank charges during the night from 03:00 to 05:00 when the price is low. The charging stops when the tank temperature reaches 90 °C, and the temperature falls during the day as there is a demand for hot water in the building. At the end of the day the water temperature reaches 55 °C and the heating element tries to maintain 55 °C by only covering the heat losses.

The energy flexibility in the hot water heaters can be realised by reducing the amount of heating elements activated simultaneously. The power input can be activated as the hot water temperature decreases below a certain self-decided threshold, or whenever necessary, if the hot water demand is known. The latter is what happens in the optimization model, as the hot water demand is input data, or a parameter in the model. The model will try to charge the hot water tank when the price is low or when there is surplus electricity produced from the PV system. If there is no surplus electricity and the price is high, the tank will try to maintain a low temperature, since the heat loss is then lower.

When looking at figure 7.3, it is noticeable that the heating element activates around 14:00, which is due to a drop in the electricity spot price. It maintains the temperature during the hour so that the tank contains enough energy to take care of the remaining demand for the day, without the need to charge during the high price hours. It is interesting to notice that the hot water tank could choose to charge at

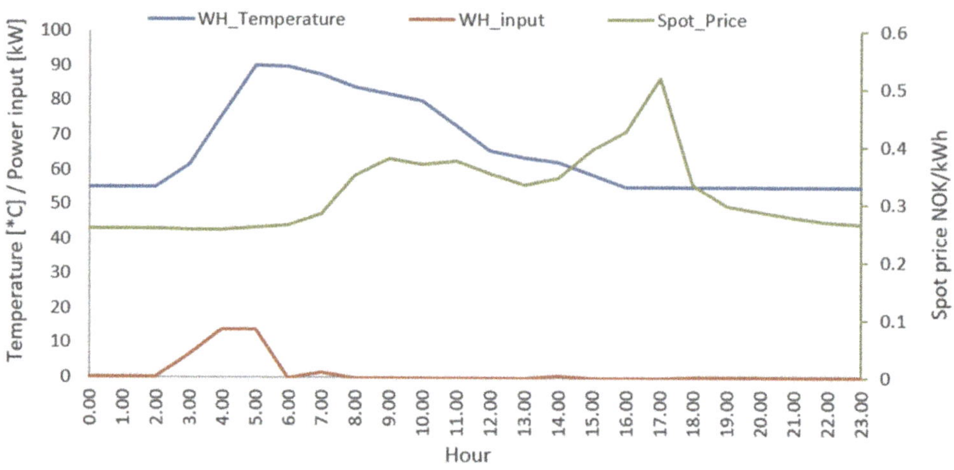

Figure 7.2. Hot water temperature and heating element power input in the subscribed power scenario, during a day with no surplus electricity and high spot price.

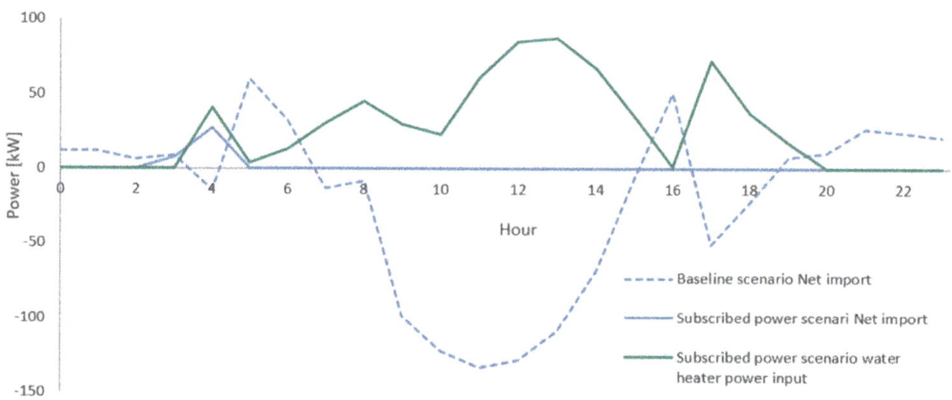

Figure 7.3. Water heater operation during a sunny day for the subscribed power scenario, compared with the net import of the baseline scenario with no flexibility.

13:00, when the price is slightly lower. However, it seems that the heat loss is significant enough, so that it is cheaper to postpone the power input one hour to save the extra heat loss from the increased temperature.

7.2.3 Operation of water heaters and batteries

In figures 7.3 and 7.4 the operation of the battery and the hot water heaters during a sunny day for the subscribed power scenario is presented.

Figure 7.4 shows that the battery discharges during the night, reducing the power import and gets ready to charge with surplus electricity during the day. For the first part of the day the battery charges and reaches a 100% state of charge (SoC) at 12:00, contributing to a net zero electricity import. It can be observed that the water heaters activate at 04:00 resulting in more import of electricity for this hour

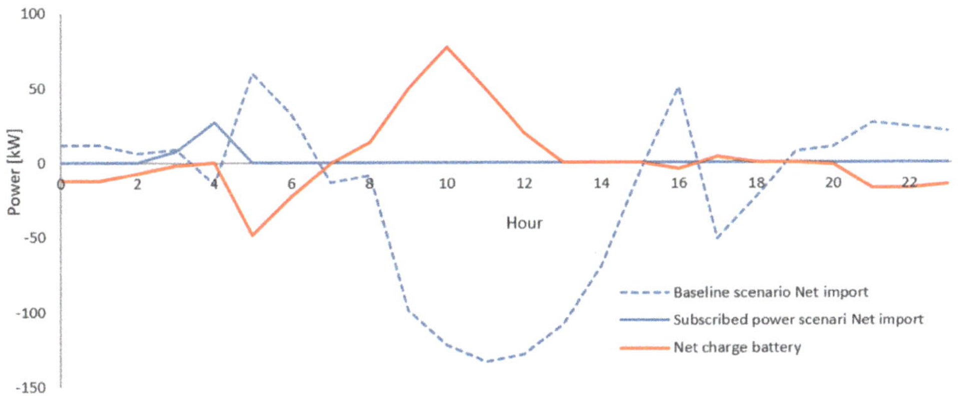

Figure 7.4. Battery operation during a sunny day for the subscribed power scenario, compared with the net import of the baseline scenario with no flexibility.

compared to the baseline scenario. However, all the water heaters reduce their power input to almost zero for the next hour. With the battery being able to provide the necessary electricity for the hour, the net import for the neighbourhood becomes zero. During the day the heating elements are active, charging close to maximum power with the surplus electricity from the PV system, keeping the net electricity import at zero.

The value of utilizing surplus energy is noticeable in these results. From figures 7.3 and 7.4, the operation of the battery and the hot water heaters for a sunny day in the subscribed power scenario can be seen. In the morning the battery discharges as the buildings start their ventilation systems, and possibly also their cooling systems. The power from the battery is most likely charged by surplus electricity from the day before. The battery also charges during the first part of the day, prioritized over the hot water heater. It discharges again during the evening. Looking at the net import of electricity in the baseline scenario, the neighbourhood consumes electricity at 16:00. In comparison to the scenario with flexibility in figures 7.3 and 7.4, the model shuts down the heating element and discharges the battery instead, avoiding the demand of electricity from the grid.

7.2.4 Peak shaving

In figure 7.5, the duration curves for the different grid tariffs are presented. The maximum power consumption, export and the total export for the different models are presented in table 7.3. From figure 7.5 it is easy to see how the different price signals affect the consumption and export of electricity. In the baseline scenario with no flexibility a typical duration curve can be seen. A steep decline from 353 kW down to around 250 kW is observed. It reaches zero at around 7400 h and maximum export is 297 kW. For the time of use (TOU) scenario, a somewhat similar trend is observed. What is interesting is that the maximum power consumption is higher than in the baseline scenario, which means that the tariff fails to reduce the peak load in the neighbourhood. The maximum consumption happens on the 21st of

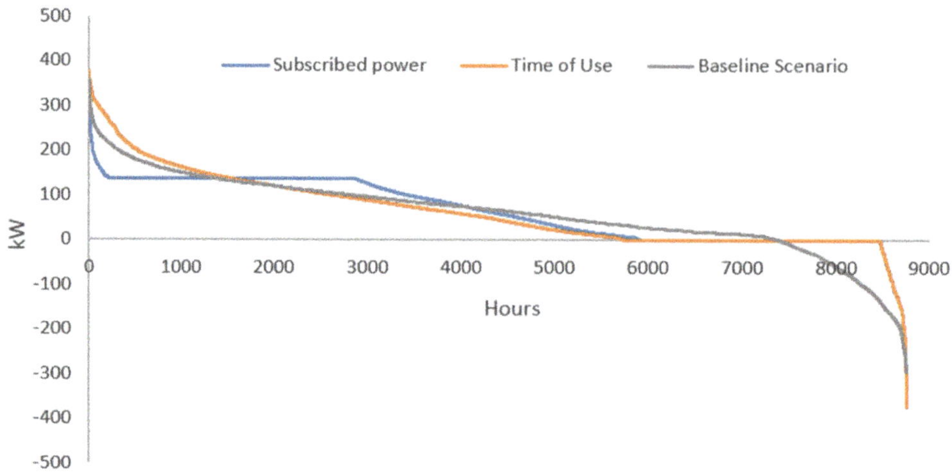

Figure 7.5. The duration curves.

Table 7.3. Maximum import and export for the different models.

	Baseline	Time of use	Subscribed power
Max. consumption [kW]	356	375	323
Max. export [kW]	297	372	372
Export total [kWh]	77 540	30 291	30 300

The three hours with the greatest import for the TOU scenario are presented in table 7.4.

Table 7.4. Time of use peak hours.

	375 kW	372 kW	370 kW
Date and time	February 21. 05:00	February 23. 02:00	February 20. 02:00
Water heater input total neighbourhood	89.95 kW	75.76 kW	89.10 kW
Battery charge	120 kW	120 kW	120 kW
Battery SOC	204 kWh	204 kWh	204 kWh
Spot price	0.2144	0.1899	0.1908
TOU price	0.1990	0.1990	0.1990

February at 05:00. The neighbourhood has a great demand for power at this hour; however, both the battery and all the hot water heaters charge at full power simultaneously. What is interesting is that for all the seven hours with consumption higher than in the baseline scenario, the consumption happens between two and five in the morning. In fact, all the 23 hours with highest consumption for the TOU scenario happen during the night-time, also between two and five in the morning, and are all during the winter. To reduce the peaks happening at night, the TOU tariff should also consider including cost of power.

The TOU tariff is designed to give price signals based on assumed peak hours in the winter. The assumed hours are between 6:00 in the morning and 8:00 in the evening. The price during these hours is 250% higher than the hours with assumed low demand. The charging of the battery and the hot water heaters contributes to 56% of the total consumption in the hour with highest consumption. This implies that in a neighbourhood with a high degree of available flexibility, the grid tariff time of use might fail to reduce the peak loads. However, the duration curve reaches zero much earlier, implying a higher degree of self-consumption. This is also seen in the way the duration curve remains around zero for approximately 2750 h.

For the subscribed power scenario, a different duration curve is observed. The maximum consumption is only 323 kW, and the decline in power is steep, reaching the optimum subscribed power threshold after around 225 h. The threshold is at 136 kW, and the duration curve remains at the threshold for around 2650 h. The optimal power threshold was found to be 136 kW. Only 1298 kWh, or 0.2% of the energy is consumed with a power higher than the threshold. The duration curve remains at the threshold for around 1900 h. The reason for this is the price signal. The tariff model is designed in a way so that electricity is less expensive when consuming below the power threshold. The price for power above the power threshold is 1 NOK per kWh per h compared to the energy part, which is only 0.04 NOK per kWh. This means that the model will utilize the flexibility sources as much as possible to stay below the power threshold.

Even though the threshold is only at 136 kW, and only 0.2% of the electricity is consumed with a power above 136 kW, the maximum power consumption is still at 323 kW. The reason is that the grid tariff does not differentiate the cost for power consumed above the threshold. Moreover, if you consume above the threshold, you pay the same for the amount of electricity, regardless of the power the electricity is consumed at. This shows that the grid tariff has a great potential to reduce the peak loads in a neighbourhood; however, the design could be improved. The cost of consumption above the power threshold could be doubled for every kW exceeding the power threshold. Increasing the cost quadratically per kW could prove to be even more effective.

The power threshold of 136 kW is only 42% of the maximum power consumption. It implies that for a neighbourhood with a high amount of storage capacity available for flexibility, a change in the model structure could significantly reduce the maximum consumption in the neighbourhood. Regardless, the tariff succeeds in reducing the maximum consumption in the neighbourhood, as the 323 kW is only 90% of the peak load in the baseline scenario.

With reference to table 7.5, 5 h with consumption over 320 kW for the baseline scenario is compared to the same hours for the subscribed power scenario. What is interesting to see in table 7.5 is that for the original hour with maximum consumption, the heating elements are turned off in the subscribed power scenario, even though the demand for hot water in the neighbourhood is 33 kW. In fact, the water heaters reduce their power inputs significantly, almost 100% for all the

Table 7.5. Peak hours for the baseline scenario compared with the subscribed power scenario.

	Baseline	Subscribed power			
Hour	Consumption [kW]	Consumption [kW]	Battery discharge [kW]	Water heater power demand [kW]	Water heater input [kW]
1278/06:00	356	323	0.0	33.3	0.0
1158/06:00	344	311	0.0	33.3	0.0
1182/06:00	343	310	0.0	33.3	0.1
1159/07:00	337	163	104.9	68.4	0.0
1279/07:00	333	267	0.0	68.4	2.2
1183/07:00	331	201	66.1	68.4	4.2

Table 7.6. Electricity export for the different models.

	Baseline	Time of use	Subscribed power
Export total [kWh]	77 540 kWh	30 291 kWh	30 300 kWh
Percentage of export of the total PV production	26.6%	10.4%	10.4%
Self-consumption	25.2%	30.7%	30.7%
Water heater energy input		28 554 kWh	28 937 kWh
Battery energy charge		21 708 kWh	21 615 kWh

original peak hours. It is interesting to notice that during the three hours with the highest power consumption, the battery does not discharge at all. However, for the fourth hour, it discharges with a rate of 104 kW, almost 90% of its maximum power. This results in reducing the power consumption from 337 kW to 232 kW. With the help of the water heaters, the resulting consumption is as low as 163 kW, only 48% of the original power consumption. The reason for this behaviour is that for the next hour, the battery discharges 94 kW, to reduce the power consumption below the power threshold which significantly reduces the cost of electricity.

Regardless of the reduced maximum consumption, the maximum export of electricity in the TOU and subscribed power scenario is even higher than the maximum consumption occurring in the baseline scenario. This means that for neighbourhoods with a high degree of self-generation from variable renewable energy (e.g. solar PV), the grid tariffs should be designed to consider the maximum export.

7.2.5 Self-consumption

As the duration curve in figure 7.5 implies, both the grid tariffs TOU and subscribed power utilizes much more of the locally produced electricity from the PV system. In table 7.6 the amount of electricity exported to the grid in the different scenarios is presented. It also presents the percentage of exported energy to the produced electricity from the PV system and the self-consumption percentage. The increased water heater power input and battery charge during export hours in the baseline scenario is also presented. In the baseline scenario, the neighbourhood exported about 77.5 MWh of energy to the grid. However, in the TOU and subscribed power scenario, the neighbourhood only exported around 30 MWh, only 39% of the baseline scenario.

The results presented in table 7.6 show that for both the grid tariffs, the energy flexible sources try to utilize more of the locally produced electricity. It results in reduced export to the grid and a higher degree of self-consumption. The results show that both the tariffs work well in increasing the self-consumption in the neighbourhood. The TOU is slightly better, however, only by a few kWh. Hot water heaters also contribute the most to the increased self-consumption in the neighbourhood, with around 57% of the reduced export in both the tariff scenarios. This is since the water heaters are modelled as slightly more efficient than the battery. All except one of the water heaters converts the imported electricity to useful energy with an efficiency of 90%. A roundtrip in the battery and two times through the inverter results in an efficiency of 87.5%. For this reason, the model will utilize the water heaters first. However, there are heat losses from the water heaters. This means that the results might look different if the hot water demand was changed. If the demand for hot water was less, the charging demand would be reduced, and the same amount of surplus electricity could not be charging the hot water heaters. The model would choose to charge the battery more, since charging the hot water heater would result in too high temperatures in the water heaters.

It is important to remember that the battery stores high-grade energy and can cover the electrical demand. The battery should therefore be the prioritized storage technology when surplus electricity is present. The hot water heaters only store heat, and the heat can potentially be provided by other energy carriers than electricity. If available, waste heat can cover most of the energy demand for hot water through a heating network.

7.2.6 CO_2 emissions

In figure 7.6, the CO_2 equivalent emissions in the Norwegian electricity grid in 2015 are presented [13]. The figure presents the average emissions for each hour of the day.

To get an understanding of the variations in the CO_2 equivalent intensity, the highest and the lowest CO_2 emissions are presented in figure 7.7. The average CO_2 emissions in the electricity network do not vary much for the hours of the day. The highest average is observed during the night at 03:00, which is 20 g CO_2 equivalent per kWh, and the lowest at 18:00 is 12.5 g CO_2 equivalent per kWh. What is interesting is that 03:00 is also the hour with the highest standard deviation. Norway imports a lot of electricity during the night since the electricity price is low. Low prices indicate a high share of residual power in the electricity grid, which could be surplus wind power. The average CO_2 emission is higher during the night than the

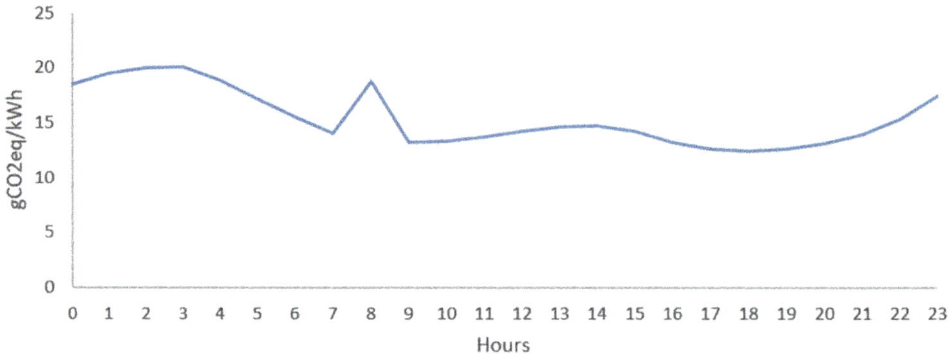

Figure 7.6. Average CO_2 equivalent emissions in the Norwegian electricity network for each hour of the day in 2015 (data from [13]).

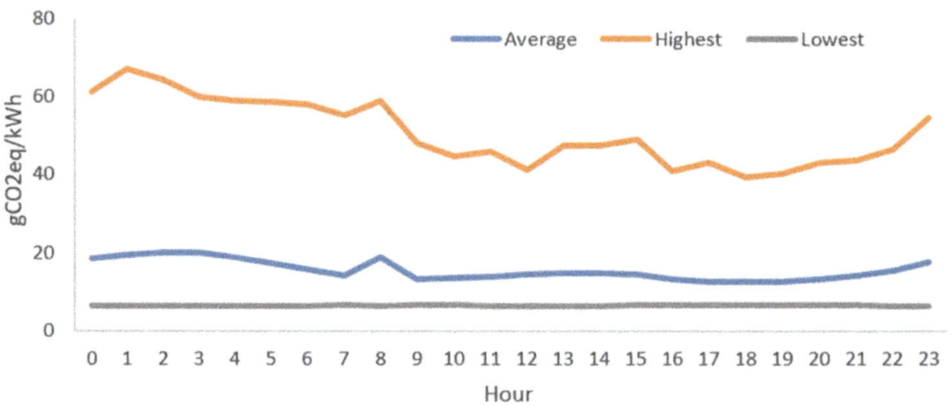

Figure 7.7. Average, highest and lowest CO_2 emissions in the Norwegian electricity network in 2015 (data from [13]).

day which implies that there is a share of non-renewables (e.g. coal power running during the night). The high standard deviation for these hours also implies that the share of renewables in the electricity mix varies. However, the highest CO_2 emissions are less than 70 g per kWh, which means that the share of renewables during the night is high for every hour.

The lowest average is observed at 18:00, which is the hour with the lowest standard deviation. This could be due to the peak power demand during the evening as people come home from work, cook dinner or take a shower. The peak power demand is usually covered with hydropower which explains both the low carbon intensity and the low standard deviation. Peak power demand usually also occurs during the morning, as people make their coffee and the office buildings start their ventilation systems.

In table 7.7 the resulting CO_2 emissions for each scenario are presented. The emissions are related to the import and export of electricity. Around one to two tons of $CO_{2_{eq}}$.emissions are saved by optimizing the control of the water heaters

Table 7.7. The CO_2 emissions related to net electricity import for different scenarios.

Baseline [kg CO_{2eq}]	Time of use [kg CO_{2eq}]	Subscribed power [kg CO_{2eq}]	Minimize CO_2 emissions [kg CO_{2eq}]
9039	10 261	9730	8003

Table 7.8. The total cost of electricity for the different grid tariffs.

Baseline scenario	Time of use	Subscribed power	Minimize CO_2
404 600 NOK/year	384 100 NOK/year	365 400 NOK/year	447 400 NOK/year

and the batteries with respect to minimizing CO_2 emissions. A reduction of one tonne, or 11% is obtained when comparing to the baseline scenario where no flexibility is present.

For the scenarios with cost optimization, the CO_2 equivalent emissions increase. When compared to the baseline scenario, the emissions increase with around 8% for the subscribed power scenario and as much as 14% for TOU. The reason is the price signal. The models optimize to reduce the costs, and as the price is lower during the night, the hot water heaters and batteries consume more during these hours. As mentioned earlier, the CO_2 intensity is in general higher during the night, which results in higher CO_2 emissions for the flexibility scenarios. The TOU also invites the system to import more during the night, as the hours from 06:00 to 20:00 during the winter are regarded as peak hours. The price is 250% higher for these hours, and as a result the model chooses to shift the load to the night and thus the CO_2 emissions increase.

7.2.7 Economic value

In table 7.8, the resulting yearly total cost for the different scenarios are presented. It becomes clear that the subscribed power tariff is favourable in reducing costs for the consumer. In the TOU scenario, the costs are also reduced, however, only 5% reduction is achieved. The reason might be the power threshold in the subscribed power scenario, as the battery and the hot water heaters are able to maintain the electricity consumption below the power threshold for 97% of the hours. The model also finds the optimal power threshold, a feature which is not present for the TOU scenario. A total of 39 200 NOK per year can be saved in the subscribed power scenario. This corresponds to a cost reduction of around 10% annually, which is significant. This is different from what Clauß *et al* (2019) found, as they did not experience cost savings for the price-based control, since the lower prices were outweighed by the increased energy use [13]. Even though the cost savings achieved in this analysis are significant, the model knows both the load profiles, the price and

Table 7.9. The cost when optimizing to reduce CO_2 emissions.

	Baseline	Minimize $CO_{2eq.}$	Difference
Emissions	9039 kg $CO_{2eq.}$	8003 kg $CO_{2eq.}$	1036 kg $CO_{2eq.}$
Cost	404 600 NOK	447 400 NOK	–42 800 NOK

the PV production, and even with an optimized control, the best grid tariff can only achieve a cost reduction of around 10%.

The price per kWh for the storage technologies is 391 NOK per kWh for the hot water heaters and 1531 NOK per kWh for the batteries. Therefore, the estimated price for the storage technologies at Campus Evenstad is 178 700 NOK for the hot water heater, and 312 400 NOK for the lithium-ion battery. Looking at the storage costs, the hot water heater is by far the less expensive storage technology. Compared to the storage costs for a battery pack in the report from BNEF, the price for the hot water heater is 26% per kWh [14]. When comparing to the real costs for residential storage systems given by suppliers in Germany, the hot water heater costs only 4% of a battery system per kWh. This implies that the hot water heater is significantly less expensive. However, it is important to remember that the battery stores electricity and can provide several other services like uninterrupted power supply (UPS). The cost of reduced CO_2 equivalent emissions when optimizing to reduce CO_2 equivalent emissions can be seen in table 7.9. The cost in NOK per ton reduced CO_2 equivalent is 41 300 NOK.

Looking at the scenario with the purpose of minimizing the CO_2 emissions, an increase of total costs is observed. The total costs are increased by around 10% compared to the baseline scenario and 22% when compared to the subscribed power scenario. However, the emissions are reduced by around 17.7% compared to the subscribed power scenario. When compared to the baseline scenario, a reduction of 11.5% is observed at a cost of 41 300 NOK per ton of CO_2 equivalent emissions reduction. Most of the measures to reduce the CO_2 equivalent emissions have been suggested to have a cost up to 2000 NOK per ton of CO_2 equivalent emissions reduction, and the marginal cost to reach the target of reduced emissions by 2020 was found to be around 1100 NOK per ton of CO_2 equivalent emissions reduction [9]. This shows that controlling the flexible source in buildings based on the CO_2 intensity in the electricity grid alone might not be the most favourable measure economically.

7.2.8 Sensitivity analysis—shadow price

The shadow prices for the TOU scenario can be seen in tables 7.10 and 7.11, while the shadow prices for the subscribed power scenario can be seen in tables 7.12 and 7.13. A one-degree change in the temperature in the water heaters corresponds to a change of 13 kWh in the total storage capacity. Increasing the power input to the water heaters by 1 kW corresponds to an increase of 11 kW for the whole campus.

Table 7.10. The shadow prices for the water heater in the TOU scenario.

55 °C–56 °C	90 °C–91 °C	Increasing power input
520.0 NOK	−332.8 NOK	−246.1 NOK
40.0 NOK $kW^{-1}h^{-1}$	−25.6 NOK $kW^{-1}h^{-1}$	−22.4 NOK kW^{-1}

Table 7.11. The shadow prices for the battery in the TOU scenario.

Increasing battery capacity	Increasing maximum charging power	Increasing maximum discharging power
−57.9 NOK $kW^{-1}h^{-1}$	−0.4 NOK kW^{-1}	−2.3 NOK kW^{-1}

Table 7.12. The shadow prices for the water heater in the subscribed power scenario.

55 to 56 °C	90 °C to 91 °C	Increasing power input
321.1 NOK	−161.4 NOK	−212.2 NOK
24.7 NOK $kW^{-1}h^{-1}$	−12.4 NOK $kW^{-1}h^{-1}$	−19.3 NOK kW^{-1}

Table 7.13. The shadow prices for the battery in the subscribed power scenario.

Increasing battery capacity	Increasing maximum charging power	Increasing maximum discharging power
−27.2 NOK $kW^{-1}h^{-1}$	−0.0 NOK/kW	−1.3 NOK kW^{-1}

The shadow prices presented in tables 7.10 and 7.12 show that for the water heater, the value of the objective function is more sensitive to the increase of the lower constraint from 55 °C to 56 °C, than the upper constraint of the water temperature operation range. One of the reasons is that the water heater would have less storage capacity available for flexibility. However, the difference in shadow price for the constraints is high. The shadow price for the lower constraint is around 40% higher than for the upper constraint in the subscribed power scenario and around 50% in the TOU scenario. This implies that increasing the temperature range by reducing the lower constraint, is more valuable than increasing the upper constraint. One of the reasons could be the heat losses; as discussed earlier the heat losses at 90 °C are twice compared to at 55 °C, if the surrounding temperature is 20 °C. Therefore, increasing the upper bounds from 90 °C to 91 °C would yield higher heat losses, and decreasing the lower constraint from 56 °C to 55 °C would yield less heat loss.

The shadow prices per kWh for the hot water heater show that the cost of increasing the lower constraint from 55 °C to 56 °C is 40 NOK per kWh in the TOU scenario and 25 NOK per kWh in the subscribed power scenario annually. The reason for the difference might be that in the TOU scenario, the price is higher during the day in the winter, between six in the morning and eight in the evening. These hours are regarded as hours when the grid is heavily loaded. This reflects in the time the shadow prices occur. For the 147 h when the shadow price is higher than 0.2 NOK, 93% of these hours occurs at eight in the evening, the last hour of the peak hours. This implies that there is great value in charging the hot water heater during the night, postponing the charging during the day, since the price is higher. However, the peak hours in Norway usually happen during the morning and the evening. If the grid tariff does not differentiate between morning, day and evening, the water heaters would activate during peak hours in the evening due to the low water temperature in the tank. Therefore, it could be reasonable to design the time of use tariff so that the hours during the day are regarded as outside peak hours. With such a tariff structure, the hot water heater can charge during the day, when the grid is not heavily loaded, decreasing the chance of the heating element activating during peak hours in the evening.

Comparing to the storage price of 391 NOK per kWh, the price of increasing the storage capacity is much higher than the annual cost of increasing the lower constraint from 55 °C to 56 °C. However, 40 NOK per kWh annually means that in around ten years the cost of increasing the storage capacity is earned back. This shows that it might be more valuable to increase the storage capacity compared to increasing the lower bounds of the water temperature range.

Looking at the shadow prices of the battery, they imply that the battery is oversized. An increase in the storage capacity will yield a slight decrease in annual costs, but compared to the cost of the battery, the savings are far from being profitable. However, it is important to remember that the battery can provide other services, such as UPS, which are not considered in this analysis. The value of increasing the power capacity, is almost zero for both the time of use and subscribed power scenario. The highest power demand in the neighbourhood in the baseline scenario is around 350 kW, which means that by discharging the battery at full power, the power import can be reduced to 217 kW, around 66%. Together with the hot water heaters, they can potentially reduce the power import to 40%. This is also shown in the duration curve, that together they keep the import below the power threshold for 97.4% of the hours. 217 kW is still well above the power threshold of 136 kW, meaning that by increasing the power of the battery, it could potentially reduce the peak even further and possibly stay below the power threshold for even more hours. However, there is no incentive for this in the grid tariff design, since the tariff does not differentiate the price for power consumed above the threshold. This is also the reason for the steepness in the duration curve, from 323 kW down to the threshold of 136 kW in only 225 h. This strengthens the hypothesis that a new design of the grid tariff, with a quadratic increase in cost for the power consumed above the threshold will result in significantly reduced peak loads in the neighbourhood.

7.3 Conclusion

The grid tariff subscribed power works best in reducing the peak loads in the neighbourhood, as the maximum consumption is reduced by 10% compared to the baseline scenario. However, if the grid tariff increases the price per kWh per h when consuming power above the power threshold, the tariff could have the potential to reduce the peak loads substantially. It is suggested to investigate a quadratic increase in the power price above the threshold. The hot water heater works best in reducing the peak load, as it does not activate during the peak hours, resulting in a reduction of between 30–70 kW for the six hours with highest consumption in the baseline scenario. The battery is mostly operating to reduce the consumption so that the neighbourhood stays below the power threshold.

The time of use tariff does not work in reducing the peak loads, as the maximum consumption is around 6% higher than in the baseline scenario. This is due to the big price difference during the winter, as the neighbourhood consumes more power outside the peak hours to reduce costs. For both the grid tariffs the maximum export is higher than the peak load in the baseline scenario. For neighbourhoods with a high degree of self-generation from variable renewable energy, such as solar PV, the grid tariffs should be designed to consider the maximum export.

Both the grid tariffs increase the carbon footprint in the neighbourhood. The model decides to consume more electricity during the night since the price is lower, however, the CO_2 intensity in the electricity grid is higher. The TOU model increases the carbon footprint the most, which is also due to the price difference in the winter, resulting in even more consumption during the night. The cost of the reduced CO_2 emissions was found to be around 41 000 NOK per ton of CO_2.

When it comes to self-consumption, both the grid tariffs work just as well as each other. The export is reduced from 77 500 kWh to around 30 300 kWh for both the tariffs. The hot water heater works best in terms of self-consumption as it stands for around 57% of the reduced export in both the tariffs. More hot water storage capacity can prove to be profitable for reduced costs in the neighbourhood. The hot water heater is also less expensive than the battery, and the battery is oversized when only considering cost savings. This analysis shows that if the purpose is to reduce peak loads, increase the self-consumption and reduce the CO_2 emissions in a ZEN, the hot water heater is the most valuable resource.

References

[1] Wiik M R K *et al* 2018 Zero Emission Neighbourhoods in Smart Cities: Definition, Key Performance Indicators and Assessment Criteria: Version 1.0. ZEN Rep.

[2] N V E 2017 Forslag til endring av forskrift om kontroll av nettvirksomhet

[3] Jensen S Ø *et al* 2017 Annex 67: energy flexible buildings-energy flexibility as a key asset in a smart building future

[4] Nuytten T, Claessens B, Paredis K, Van Bael J and Six D 2013 Flexibility of a combined heat and power system with thermal energy storage for district heating *Appl. Energy.* **104** 583–91

[5] De Coninck R and Helsen L 2013 Bottom-up quantification of the flexibility potential of buildings *Proc. of BS 2013: 13th Conf. of the Int. Building Performance Simulation Association* pp 3250–58
[6] BNEF 2018 Energy storage is a $620 billion industry
[7] Goldie-Scot L 2019 A behind the scenes take on lithium-ion battery prices *Bloomberg NEF*
[8] IRENA 2017 Electricity storage costs
[9] Klimakur 2010 Tiltak og virkemidler for å nå norske klimamål mot 2020. TA2590/2010, Regjeringen
[10] Fan J 2009 System for domestic hot water services. educational notes for 11127 sustainable heating and cooling of buildings
[11] Lienhard J H 2011 *A Heat Transfer Textbook* (Chelmsford, MA: Courier Corporation)
[12] Von Meier A 2006 *Electric Power Systems: a Conceptual Introduction* (New York: Wiley)
[13] Clauß J, Stinner S, Sartori I and Georges L 2019 Predictive rule-based control to activate the energy flexibility of Norwegian residential buildings: case of an air-source heat pump and direct electric heating *Appl. Energy.* **237** 500–18
[14] BNEF 2019 A behind the scenes take on Lithium-ion battery prices [Internet] [cited 2019 May 21]. Available from: https://about.bnef.com/blog/behind-scenes-take-lithium-ion-battery-prices/

IOP Publishing

Energy-Smart Buildings
Design, construction and monitoring of buildings for improved energy efficiency
Jacob J Lamb and Bruno G Pollet

Chapter 8

Spot price and carbon emissions in a zero-emission neighbourhood

Anne A Brajkovic, Mathias Bjugan, Jacob J Lamb, Inger Andresen and Kristian M Lien

To reduce global warming, greenhouse gas emissions from the building sector need to be reduced. This can be done by establishing zero-emission neighbourhoods. Campus Evenstad in Norway is a pilot project with the goal to develop into such a neighbourhood. In a zero-emission neighbourhood, all emissions come under scrutiny. To reach zero emission, it is necessary to produce clean energy locally. This chapter gives a detailed case study of the carbon emissions of Campus Evenstad in relation to the cost and source of energy used.

8.1 Introduction

To mitigate climate change, greenhouse gas emissions from the building sector need to be reduced significantly. One way to achieve this is by establishing zero-emission neighbourhoods. For neighbourhoods to reach zero emissions, it is necessary to produce clean energy locally, reducing the potential emissions from transporting the energy long distances. Campus Evenstad in Norway is a pilot project with the goal to develop into such a neighbourhood. Here, photovoltaic (PV) panels have been installed at the campus for local clean energy production. Although these panels can produce large amounts of energy, their source of energy (i.e. the Sun) is intermittent, so energy can only be produced periodically. Therefore, to utilize the clean energy outside of the time frame it is produced, energy storage solutions are necessary.

8.1.1 Campus Evenstad

The photovoltaics (PVs) installed on campus Evenstad contribute to the goal of reaching a zero-emission neighbourhood by producing clean energy locally. Despite this, the energy produced from the PVs is only accessible during the day when there is the required sunlight. If PVs are used on their own, the energy that is not used

doi:10.1088/978-0-7503-3259-0ch8

during daytime is either exported to the grid for minimal profit or is lost. To utilize the energy from the Sun outside of the time frame it is available, sufficient energy storage is necessary.

To facilitate the requirement of energy storage, a lithium–nickel–manganese–cobalt-oxide (NMC) battery bank has also been installed at the Evenstad campus. Out of the installed capacity of 204 kW h/120 kW, 100 kWh is reserved for uninterruptible power supply (UPS) for times when the power grid is down. The remaining capacity can be used freely for other purposes. The initial cost of the battery was 2.5 million NOK, and it is assumed to have embodied emissions of 35 088 kg of CO_2-equivalents [1]. Despite this, by charging the batteries with clean energy and using the batteries when the grid delivers carbon intensive energy, the battery can compensate for 18% of the embodied emissions of the battery bank.

8.1.2 Choice of battery

Lead acid batteries are already extensively used in combination with PVs. The lifetime and depth of discharge (DoD) of this technology is short. Lithium ion batteries have shown more promising lifetime and DoD with the economic cost of production continuing to fall. Several lithium technologies exist, and they all bring different properties. NMC and lithium–manganese-oxide (LMO) contain manganese, which creates a high initial efficiency but also a shorter battery lifetime. Cobalt makes lithium–nickel–cobalt–aluminium oxide (NCA) a slightly more expensive cathode material. Both nickel and cobalt have been found to have higher environmental impacts than alternative cathode materials, in addition to environmental and exposure risks. Out of all lithium battery technologies, lithium–iron-phosphate (LFP) is the least environmentally intensive, but its low energy density can result in higher use phase emissions. The NMC batteries currently installed at Evenstad have a higher energy density and lower use phase emissions; however, they are neither the cheapest nor the most environmentally friendly with respect to production [1, 2].

Even though batteries are associated with the zero-emission solution, the emissions connected to production are high. Lithium mining in Latin America leads to water shortage and toxic spills, while nickel mining results in severely polluted waterways in Russia. Cobalt mining in DR Kongo results in dust, fumes and large amounts of wastewater, and the cell production has contributed to air pollution in East Asia [3]. By conducting a life cycle assessment (LCA), it has been shown that the production of a 26.6 kWh NMC traction battery has an embodied GHG-emission per kWh of 172 kg [2]. Of these emissions, 60% is due to a carbon intensive electricity mix used for the energy-intensive production methods. Therefore, by switching to a low carbon electricity mix, embodied emissions will be reduced. Kim *et al* [4] showed slightly lower embodied GHG-emission per kWh of 140 kg for a 24 kWh traction battery. This is most likely due to the fused cell chemistry of NMC/LMO, which contains significantly less cobalt. It must be noted that these values represent only the battery production and not the required control systems and inverters.

By using a repurposed battery, the climate impact can be significantly reduced. When an electric vehicle (EV) has run for approximately 200 000 km, its battery capacity will typically have reduced to 80% of its original capacity and is no longer ideal for its original purpose. The battery is then expected to be recycled. Battery recycling is an expensive process and only 5% of lithium-ion batteries are recycled in Europe [3]. To delay the recycling process, old battery cells can be reassembled into new batteries, as the residual capacity of traction batteries can be valuable for residential purposes. This type of battery is called a refurbished battery or a repurposed battery.

There are several barriers to repurposed batteries, but these challenges may be overcome by new battery system designs [5]. The challenges are absence of a clear framework for battery reuse, technical barriers due to lack of performance and degeneration data, economic barriers and safety barriers. Reuse of batteries follows a waste management hierarchy; which values reuse above recycling. By maintaining a product's value, the amount of waste can be reduced, resulting in lower GHG-emissions over the product's lifetime. Vehicle batteries are classified as waste after completing their usage life in an electric vehicle, according to European law. Therefore, refurbished batteries can be assumed to have zero GHG-emissions, as their production emissions are connected to their primary use. Furthermore, examinations of using LMO/NMC batteries from a plug-in hybrid EV with initial capacity of 11.4 kWh for energy storage from PVs have been performed, allocating all the impacts of the battery production and the end of life treatment (EOL) entirely to the batteries' primary use [5].

Other studies have also showed this trend. An NMC traction battery with an initial capacity of 24 kWh has been shown to reduce climate impacts if used for peak shaving or load shifting purposes in France [6]. Here, the battery was charged at night, when the contribution of renewable energy sources was generally higher with lower GHG-emissions, storing the energy for use during the day. It has been suggested that an electricity mix with a lower carbon intensity during the night could prove environmentally beneficial even when considering the efficiency loss in the battery [6, 7]. These observations suggest that there are environmental benefits when using a repurposed battery in combination with PVs, as long as the emissions from the grid electricity varies sufficiently throughout the day. The objective of this study is centred around the following research questions with reference to the environment and the economics.

- What is the level of the GHG-emissions of specific battery technologies and what is the source of these emissions?
- How can batteries help reduce GHG-emissions in a zero-emission neighbourhood?
- How much can the power tariff be reduced by peak shaving?
- How can the batteries be used to maximize the benefits, both economically and environmentally?

To find the emission variations in Norway, the power market needs to be examined closer.

8.1.3 The Norwegian power market

The Norwegian electricity market is split into two parts. One deals with the physical electricity, which is a mix from different origins, where the amount of GHG-emissions per unit of electricity rises at night when the country imports power from the continent. The other market deals with the rights to the produced electricity. A producer of renewable energy can sell certificates of origin equivalent to the amount of renewable energy produced. The consumer will then, in theory, only use electricity with low carbon intensity [8].

According to Hovland [9], 80%–90% of Norwegian certificates of origin are sold to the greater Europe. Therefore, in theory, the Norwegian grid could be argued to have an average carbon intensity of 520 g per kWh, instead of the 18.9 g per kWh that is represented by the physical mix in 2018. The European average electricity mix had a carbon intensity of 345 g per kWh in 2017 [8]. Norwegian consumers and industry refer to the physical energy produced in the country, but this practice means that on a global scale the production of clean energy in Norway is double counted. To avoid this problem, this paper will focus on the physical electricity in a region.

Most renewable energy technologies are difficult to regulate and predict, as the environment is impossible to control. Norway is in a unique position with large amounts of stored hydropower that is relatively easy and inexpensive to regulate. It has been suggested that hydropower plants will generate electric energy when the electricity prices are highest, as this brings the most economic profit [10]. As a result, the carbon intensity and the economic cost of Norwegian electricity has an inverse relationship, as shown in figure 8.1. This is a contrast to Europe, where unregulated sun power and wind power will push prices down due to the low marginal production costs for these technologies.

The fluctuations in carbon intensity and cost of grid electricity can be exploited for profit. This can be achieved either by charging the battery when the grid emissions are low and discharge when they are high. It is also possible to exploit fluctuations in spot price the same way. However, it has been suggested that residential batteries in Norway are not economically feasible [11]. Despite this, the possibility to charge the battery with low cost electricity from the grid and use it when the prices are higher has not yet been fully explored and potentially offers a different perspective. As the spot price and carbon intensity of Norwegian grid electricity is inversely correlated, it is not possible to exploit fluctuations in both economic cost and GHG emissions at the same time. Therefore, reducing the GHG emissions will have an economic cost. This is called the abatement cost and is generally standardised to the expense required to reduce emissions by one ton of CO_2 equivalents. The abatement cost can vary significantly between thousands of NOK to negative values. As an example, the cost of carbon capture and storage varies between 1300 and 2250 NOK per ton and electrification of electric vehicles costs approximately 1230 NOK per ton [12].

Another way of exploiting a stationary battery for financial savings is by peak shaving. This is done by reducing the electricity demand from the grid during cost peaks, reducing the cost of the overall electricity bill. The battery can compensate for

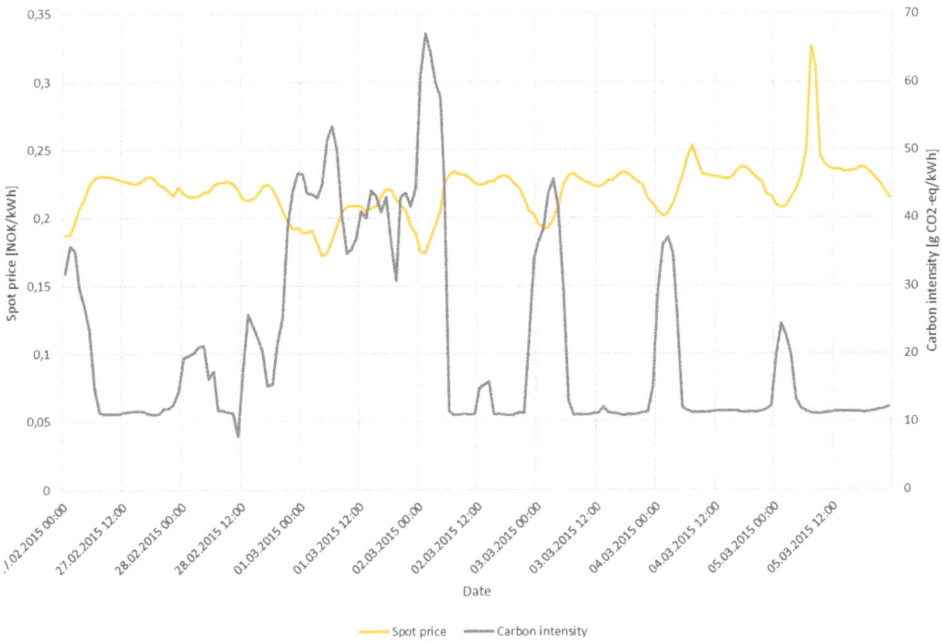

Figure 8.1. Comparison of spot price and carbon intensity in bidding zone NO1 in the period 27 February to 5 February 2015.

this by discharging when the power cost is high and charging when the power cost is low. It is also possible to save on both economic cost and GHG emissions by maximizing the self-consumption of PV-generated electricity through charging of the battery.

The value by either maximizing economical or environmental gain can be found by producing optimization models that use the available data to intelligently control the battery. As the first step to a zero-emission neighbourhood is to reduce energy requirement, the calculations are conducted in different scenarios. The currently installed PVs generate 62 000 kWh per year. The carbon debt of the installed PVs has approximately 50 g per kWh under east-Norwegian conditions. Furthermore, there is the possibility to upgrade the PVs and the current building stock at campus. This creates two potential future scenarios: a hypothetical upscaled PV (UPV) scenario, which generates 259 000 kWh per year, and hypothetical upscaled PV in addition to upgraded building stock to TEK-17 standard (UPV-TEK17) scenario, which generates 305 000 kWh per year [13].

8.2 Methodology

A number of models were developed to identify the characteristics to the battery during different scenarios. They are all linear optimization problems solved by the simplex method and modelled in Excel. The Solver (an add-in program in Excel), was used to solve linear problems. The target function, the variables, the restrictions and the solution method were all added to The Solver. The models apply historical data and have a perfect foundation for the calculations and are used to calculate the maximum value of the specific characteristic each of the applications can yield if isolated. It is not possible to merge the results from the models, as the models influence each other's outcome.

There are some assumptions associated with the calculations. The representative year used to explore savings in spot price and reductions in carbon intensity from the electricity in the models is 2015. For peak shaving and self-consumption, data from 2015 to 2017 is utilized. All calculations consider a depth of discharge (DoD) of the batteries of 80% (the recommended maximum DoD to extend the battery life), and 100 kWh of the battery capacity reserved for UPS (the battery's original purpose). This means that today's battery rarely discharges more than 51% of the full capacity. Nominal efficiency for a cycle is 96% for the battery and also for the inverter.

Degeneration of the efficiency over time is not accounted for, even though it may prove substantial. Further, it is assumed that charging and discharging the battery is based on the limits of the inverter, making the rate of discharging a bit higher than the rate of charging. The models are also based on the assumption that import and export is constant during the time period. The three first models are based on scenario CPV, while the self-consumption model utilizes scenario UPV and UPV-TEK17. All of the models share the same three variables, except peak shaving, which has two extras. Each model has five or six restrictions, whereas four is common for all.

Data per hour for carbon intensity in the electricity in bidding zone NO1 in 2015 was provided by J Clauß, while S Backe has provided hourly data for import and export of electricity to the campus in 2015 to 2017. Data for spot price on electricity in NO1 is from Nordpool [14].

8.2.1 The peak shaving model

The first model aims to minimize the cost of the power tariff by minimizing the peak shaving threshold for a given period. Campus Evenstad is settled for the highest effect peak in the last 12 months; therefore, a year is a natural boundary in the model. The peak shaving threshold and power output of the battery during maximum import are given in kWh per h, while the other variables are given in kWh. Furthermore, the target function is defined in equation (8.1), which aims to find the optimal peak shaving threshold.

$$Min\ I_{max} - xu_{max} \cdot \eta_u = t \tag{8.1}$$

Here *Min* represents that the linear problem should be calculated to the minimum possible value, I_{max} is the maximum imported electricity per hour during the time period, xu_{max} is the power output of the battery per hour with maximum energy input, η_u is the nominal discharge efficiency and t is the threshold for peak shaving. A number of restrictions follow the target function. Equation (8.2) indicates the amount of energy on the battery at the start of the next hour.

$$z_{i+1} = z_i + xi_i - xu_i \tag{8.2}$$

Here z_i is the amount of energy in the battery per unit time (i), xi_i is the energy input per unit time (i) into the battery and xu_i is the energy output per unit time (i) from the battery. Therefore, the amount of energy on the battery at the start of the next hour is the amount of energy on the battery at the start of the current hour added with the difference of the incoming and outgoing energy in the hour in question. The amount of energy in the battery also has a limitation and it is given by equation (8.3), which indicates the upper and lower limit of the amount of energy in the battery bank.

$$\frac{100\,kWh}{\eta_u} \leqslant z_{i+1} \leqslant \frac{e}{\eta_u} \tag{8.3}$$

The amount of energy must always be greater than 100 kWh, which is the amount of energy reserved for UPS. Loss of energy through nominal efficiency is included. On the other hand, the amount of energy in the battery can never be higher than the storage capacity, including the rated efficiency. Equation (8.4) indicates the upper limit for power input.

$$0 \leqslant xi \leqslant p_{inn} \cdot \eta_i \tag{8.4}$$

The power input into the battery must be between zero and the maximum incoming power (including any losses). Equation (8.5) gives the upper limit for power discharged from the battery.

$$0 \leqslant xu \leqslant \frac{p_{ut}}{\eta_u} \tag{8.5}$$

The power out of the battery must be between zero and the maximum power out. Losses are included through nominal efficiency. As shown in equation (8.6), it appears that real imports are the maximum threshold for peak shaving.

$$I - xu \cdot \eta_u \leqslant t \tag{8.6}$$

Total imports to campus minus energy out of the battery, including losses, must be the maximum peak shaving threshold.

The model works so that the battery is drained continuously as long as the consumption of power on the Evenstad campus is higher than the threshold for peak shaving. When the consumption is lower than the threshold, the battery is charged. Financial gain can be found by multiplying the power tariff (NOK) by peak shaving (kWh per h).

8.2.2 Carbon intensity model

A model has been developed to calculate the GHG emissions by using the battery. Using the same data on imported power as used in the previous model and its associated carbon intensity, the optimal usage for the battery could be calculated. The objective function is to maximize the saving of greenhouse gas emissions and is given in equation (8.7).

$$Max\sum_{i=1}^{x} xu_i \cdot \eta_u \cdot C_i - \sum_{i=1}^{x} xi_i/\eta_i \cdot C_i \tag{8.7}$$

The first step represents the energy out of the battery multiplied by the carbon intensity of imported power during the same hour. The second step is the energy entering the battery multiplied by the carbon intensity of the imported current during the same hour. Losses through nominal efficiency are included in the calculations. The goal function starts in hour 1 and sums up the results of cumulative saved emissions. The restrictions of this model are given in equations (8.2)–(8.6) (of which equation (8.6) is used only when the peak shaving threshold is intended to restrict imports), and equation (8.8).

$$xu_i \cdot \eta_u \leqslant I_i \tag{8.8}$$

Equation (8.8) shows that the battery cannot supply more energy than the need for power on campus including loss. The model specifies the optimal solution for charging and discharging the battery. Moreover, the battery is charged when the carbon intensity of the current is low and discharges when the carbon intensity of the current is high.

8.2.3 Spot price model

An additional model has been created to minimize the spot price in the electricity cost. This model is very similar to the previous one; however, the objective function is different and aims to maximize the financial profit by cutting the expenditure to the spot price point. The optimum solution specifies when to charge and discharge the battery.

$$Max\sum_{i=1}^{x} xu_i \cdot \eta_u \cdot S_i - \sum_{i=1}^{x} xu_i/\eta_u \cdot S_i \tag{8.9}$$

The function is the difference between the energy output of the battery per hour multiplied by the spot price in the current hour and the energy into the battery per hour multiplied by the corresponding spot price.

8.2.4 Consumption model

A model has also been developed to maximize the consumption of the surplus current produced by the PVs by storing it in the battery. The reason for this is that the surplus electricity that is stored for later consumption is much more economically profitable than exporting it. In order to find the potential, an optimization model has been created focusing on scenario UPV and scenario UPV-TEK17, which also deals with upgraded building stock. Time data on imported power and production from the PV system was used for this model. The goal function is to maximize consumption of the surplus stream to avoid exporting the power. The function is given in equation (8.10).

$$Max\sum_{i=1}^{x} xu_i \cdot \eta_u \tag{8.10}$$

The function states that the energy out of the battery per hour multiplied by the efficiency should be maximized. This is summed up to calculate the potential cumulative consumption of the surplus stream. The restrictions are given in equations (8.2)–(8.5) and equation (8.8). The model works on the assumption that the battery only recharges with surplus power from the PV system and discharges as soon as the system does not meet the power requirement on campus.

8.3 Results

Both new batteries and refurbished batteries are fit for the purpose of making the Evenstad campus a zero-emission neighbourhood. As the seller of the currently installed batteries on campus Evenstad did not provide emission data for production, Ellingsen's results of 172 kg per kWh are used as basis [1]. As the installed capacity is 204 kWh, the total emissions for production of this battery are calculated at 35 088 kg. Ellingsen also shows that the GHG emissions of batteries produced on a low carbon electricity mix are 60% lower, resulting in a total of 69 kg per kWh if low carbon energy is used during production. It is further assumed that with refurbished batteries, the GHG emissions are only 25%, resulting in emissions of 43 kg per kWh [1].

Estimated values based on energy savings, export, peak shaving and self-consumption of surplus PV energy for Campus Evenstad were calculated. Energy savings consist in respective order of the mean spot price, the mean marginal losses for the year, the consumption tax, the green certificate tax and 25% value added tax (VAT). Exported energy shares three segments with energy saving, in addition to an income following exporting energy to the grid. The value of yearly highest peak shaving is estimated power tariff for 2019 and 25% VAT added. The value of self-consumption of surplus power consists of the difference between imported and exported electricity, including the losses in the battery and inverter. Campus Evenstad saves on average 0.61 NOK per kWh by reducing the consumption of electricity, and the value of exporting surplus energy to the grid is on average 0.36 NOK per kWh. The value of the yearly highest peak shaving is 540 NOK per kWh per h and the value of caching excess energy from the PV system to the battery has a mean value of 0.20 NOK per kWh.

8.3.1 Peak shaving

The potential for peak shaving in scenario CPV is given in table 8.1. The threshold for peak shaving is different each year. The mean value for 2015–17 is 409.9 kWh per h. Profit from the saved power tariff can then be calculated as 27 699 NOK yearly. On average the battery is used for 8 h of peak shaving yearly. Therefore, it is possible to save 169.3 NOK per kWh per year, which can be calculated as 2534 NOK over the battery's expected lifetime (15 years). This does not support the minimum required investment cost of 6000 NOK per kWh of battery. The shadow

Table 8.1. Possibilities for peak shaving in scenario CPV.

	2015	2016	2017	Average
Three highest power peaks ($kWh\ h^{-1}$)	15/12 @ 9–10 h: 436.3 15/12 @ 8–9 h: 422.6 15/12 @ 11–12 h: 366.2	07/03 @ 11–12 h: 436.3 07/03 @ 12–13 h: 422.6 30/11 @ 10–11 h: 366.2	22/11 @ 12–13 h: 436.3 22/11 @ 12–13 h: 422.6 22/11 @ 13–14 h: 366.2	
Peak shaving threshold ($kWh\ h^{-1}$)	377.5	421.2	430.9	409.9
Peak shaving ($kWh\ h^{-1}$)	58.83	58.26	36.80	51.30
Saved power tariff (NOK)	31 766	31 459	19 872	27 699
Hours with peak shaving	4	13	7	8
Shadow price storage capacity ($kWh\ h^{-1}$)/NOK	0.4703/254	0.1568/84.65	0.3135/169.3	0.3135/169.3
Shadow price power capacity	0	0	0	0
Full utilisation of power capacity (kWh)	400.9	685.3	600.5	562.2

Data is missing between 1 January and 8 February 2017.

Table 8.2. Value of peak shaving in scenario UPV.

	2015	2016	2017	Average
Peak shaving ($kWh\ h^{-1}$)	58.83	31.08	41.93	43.95
Peak shaving threshold ($kWh\ h^{-1}$)	377.5	404.2	425.7	402.4
Saved power tariff (NOK)	31 766	16 785	22 644	23 732

Data is missing between 1 January and 8 February 2017.

price for power capacity is zero as it is not possible to peak shave more as in this case the battery size is the limiting factor. The last row in the table shows the required size of the battery to fully exploit the capacity of 120 kW, which is on average 562.2 kWh. Theoretical maximum saving by peak shaving with a battery of this size has an economical value of 56 160 NOK in yearly power tariffs saved.

Results for peak shaving in scenario UPV are given in table 8.2. The table shows that on average 43.95 kWh per h is peak shaved. This is 7.35 kWh per h less than for scenario CPV and makes a saving of 23 732 NOK per year. In scenario UPV-TEK17, the threshold for peak shaving is 264.3 kWh per h. The following peak shaving average is 81.6 kWh per h with a yearly profit from saved power tariffs of 44 060 NOK.

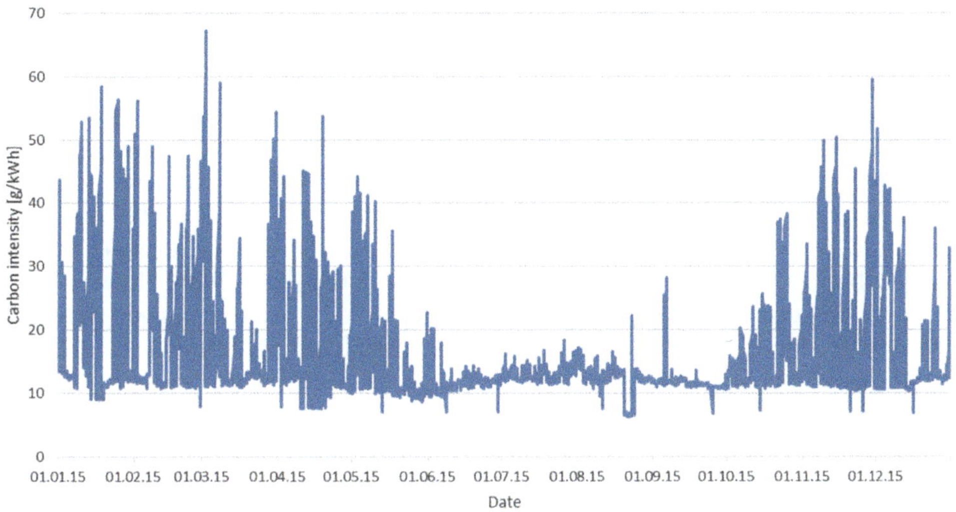

Figure 8.2. Carbon intensity of grid energy in bidding zone NO1 in 2015.

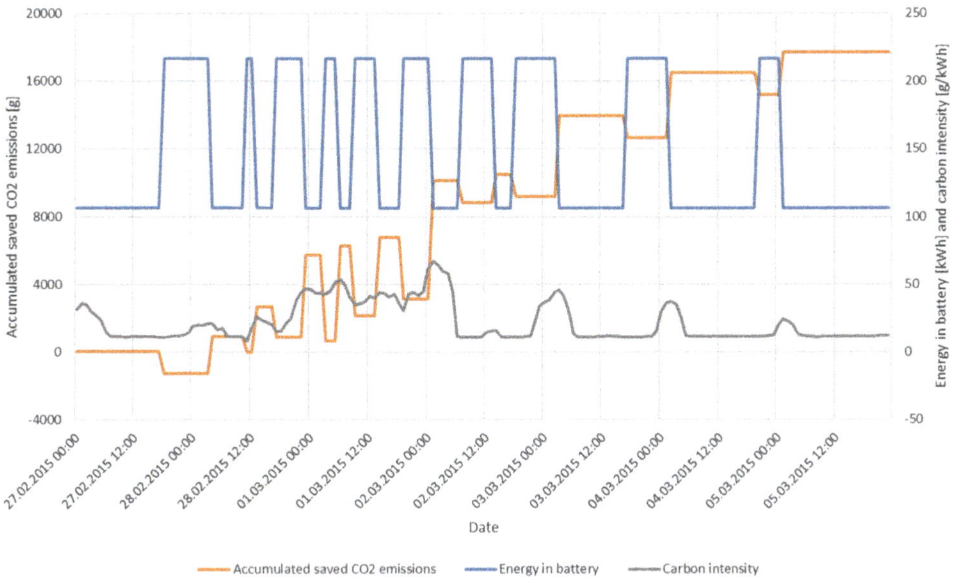

Figure 8.3. Fluctuations in carbon intensity is utilized to save CO_2 emissions in the period 27 February to 5 March 2015.

8.3.2 Carbon intensity

The carbon intensity of imported electricity in bidding zone NO1 is shown in figure 8.2. The intensity alternates between 6.2 and 67 g of CO_2 equivalents per kWh. The lowest level is found on the 22nd of August between midnight and 1 am; whereas, the largest level is found on the 2nd of March between 1 and 2 am. The carbon intensity is generally higher during winter and night-time. According to available data, the average CO_2 equivalents-intensity was 15.2 g per kWh in 2015.

Figure 8.3 shows a week in 2015 where the fluctuations in carbon intensity are used to reduce GHG emissions. The model showed a possibility to save 18 kg of CO_2 equivalent emissions during this week. The average for the rest of the year is less, as this week offered especially large alternations in GHG emission intensity. Throughout 2015 there was the possibility of saving 410 kg of GHG emissions. This could be as high as 6150 kg over the lifetime of the battery. Despite this, the installed battery with a GHG emission debt of 35 088 kg will still have a GHG emission debt, even if the entire battery capacity could be used for carbon reduction. A battery produced on low carbon energy or a refurbished battery will also still have a GHG emission debt for their respected embodied emissions, though it will be significantly less.

8.3.3 Spot price

The spot price on electricity in bidding zone NO1 between 2015 and 2017 is presented in figure 8.4. The mean spot price during this period was 0.23 NOK per kWh, with the minimum spot price of 0.01 NOK per kWh taking place on the 26 July 2015 between 4 and 5 am. Due to increased demand, the prices are higher during winter. Therefore, it is no surprise that the maximum spot price of 2.08 NOK per kWh occurred on the 21 January 2016 between 7 am and 8 am.

The economic gain of exploiting the price difference one week in 2016 with exceptionally high alternations in spot price is given in figure 8.5. The profit for this week is 710 NOK. The average gain over the year will be lower, as the average week has less alternations in the spot price. Overall, the battery can save 1765 NOK over a year, totalling 26 475 NOK over the battery's lifetime.

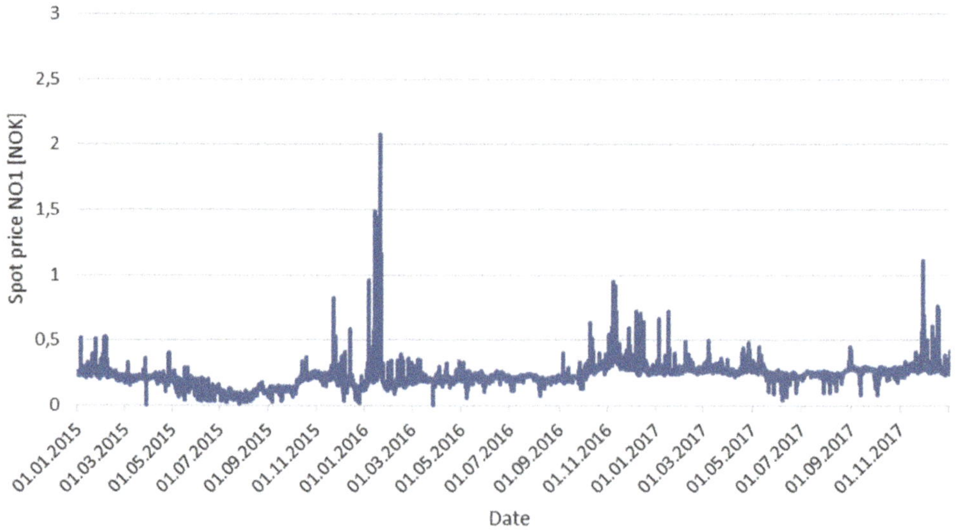

Figure 8.4. Spot price in NO1 from 2015 to 2017.

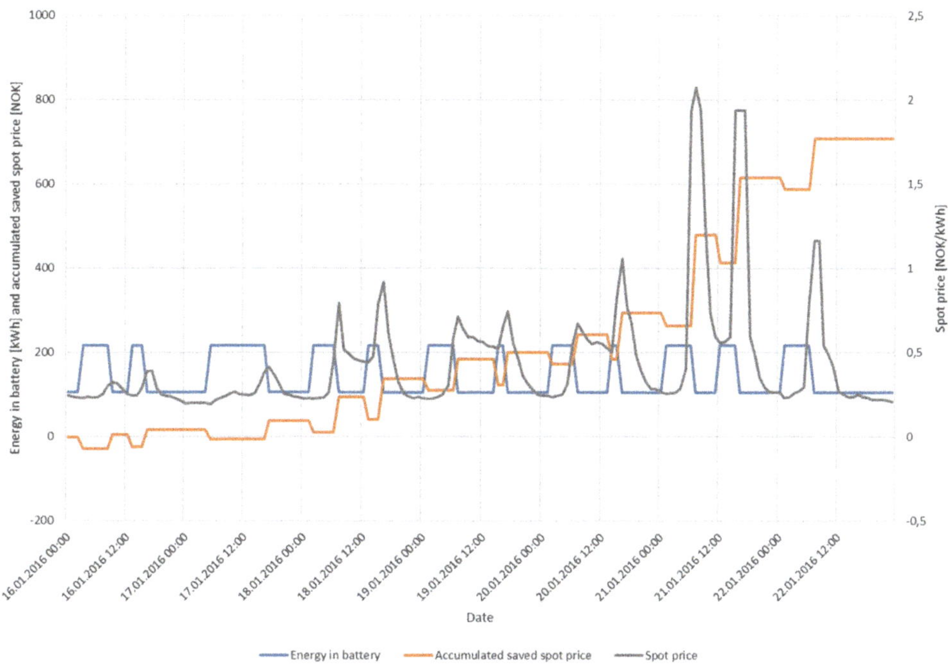

Figure 8.5. Fluctuations in spot price are utilized to make profit in the period 16 January to 22 January.

8.3.4 Comparison between carbon intensity and spot price

As the spot price and carbon intensity of the electricity are inversely correlated in the Norwegian power market, it is extremely difficult to generate a financial profit while reducing the GHG emissions. Figure 8.6. shows the accumulated saved GHG emissions for a chosen week, which results in a reduction of 18 kg CO_2 equivalents. This leads to a cost increase of 48 NOK. Therefore, the cost is 2670 NOK per ton reduction in GHG emissions. This can be compared to the abatement cost for carbon capture and storage, which varies between 1300 and 2250 NOK per ton, and electrification of electric vehicles, at approximately 1230 NOK per ton. Over the battery's presumed lifetime, the cost will be 16 575 NOK to save 6208 kg GHG emissions.

By maximising the economic benefits, the financial gain will be 23 NOK, with an increase of 19 kg of GHG emissions (figure 8.7). Therefore, there is a production of 1.2 kg of GHG emissions per NOK saved. By normalizing this cost for the battery's entire lifetime, 26 475 NOK will be saved with an increase in GHG emissions of 31 770 kg.

8.3.5 Self-consumption

The value of saving excess energy in a battery instead of exporting it to the grid in scenario UPV is about 2500 NOK per year, which corresponds to an increase in self-consumption of 12 500 kWh per year. The condition that the battery is used only to

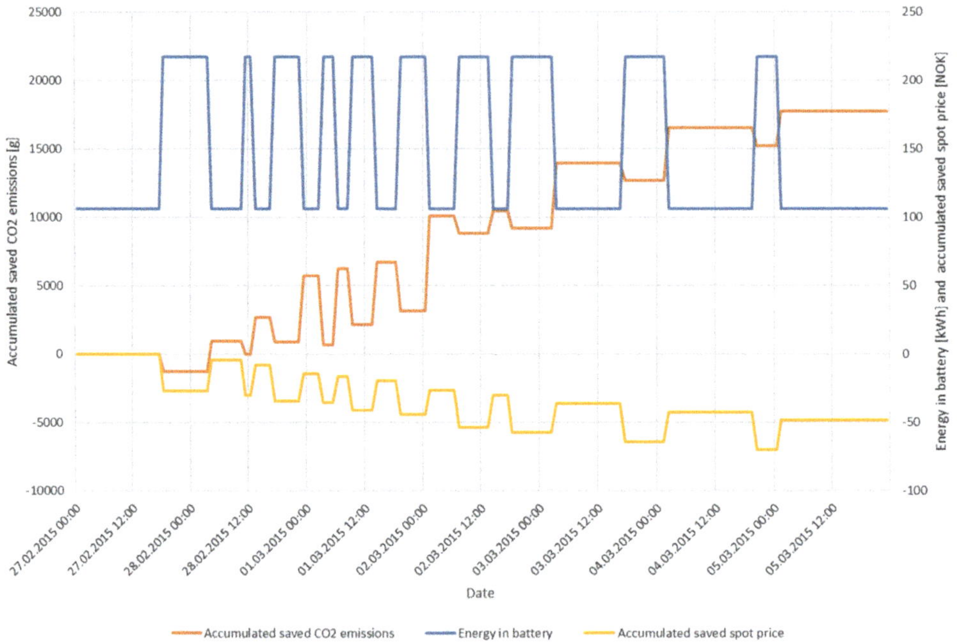

Figure 8.6. Maximized accumulated saved GHG emissions for the period 27 February–5 March 2015.

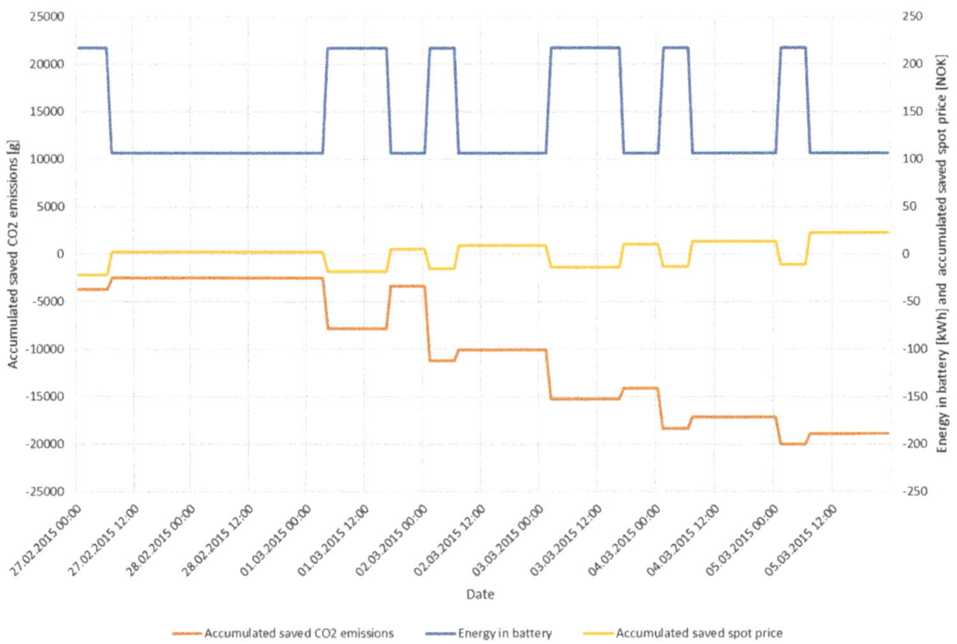

Figure 8.7. Maximized accumulated saved spot price for the period 27 February–5 March 2015.

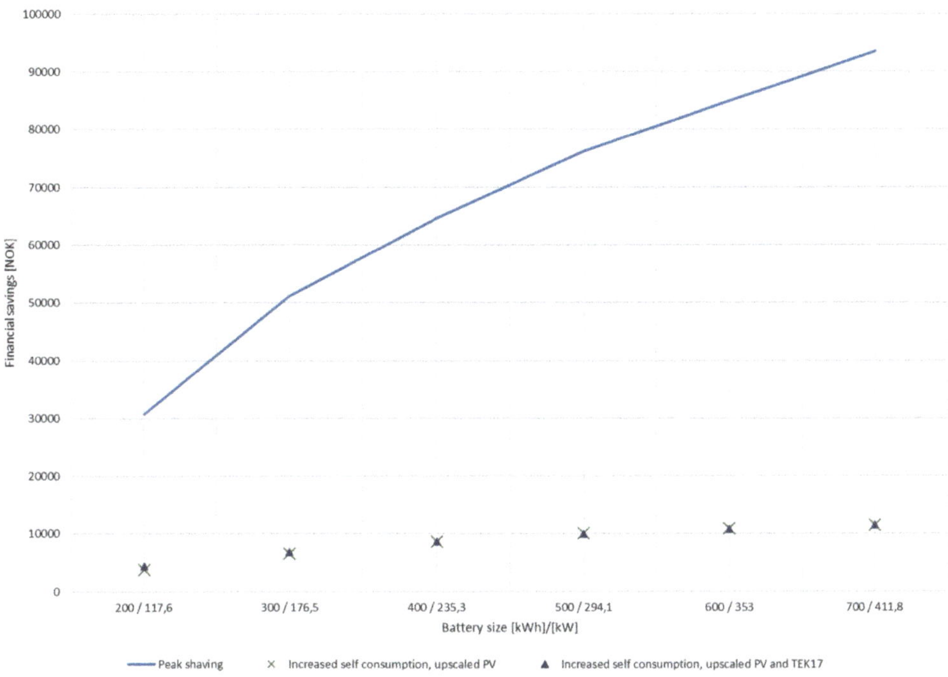

Figure 8.8. Comparison of various battery sizes for use on campus Evenstad in 2015.

store surplus energy and supply the campus as soon as it needs energy applies. The campus then saves 0.20 NOK per kWh by storing PV energy in the energy instead of exporting it. In scenario UPV-TEK17, increased self-consumption will be 14 190 kWh per year, with economical savings of 2848 NOK per year. The environmental benefit of increased self-consumption is negligible, as the PV system on campus generally leads to a much higher carbon intensity than the power grid. By exporting the excess energy, losses in the battery and inverter are also avoided.

8.3.6 Size of the battery bank

The economical value of peak shaving and increased self-consumption of PV-energy for different battery sizes in 2015 is shown in figure 8.8. For a 300 kWh battery the value of peak shaving, increased self-consumption and in addition maximised saved spot price is 2852 NOK per kWh over the battery's lifetime, which is smaller than the investment cost, which is at least 6000 NOK per kWh. Since the effect of upgrading the battery size decreases gradually from 300 kWh to 700 kWh and the price is estimated to be linear, there is no economic gain in upgrading the battery size.

GHG emission reduction and financial savings when the highest import from the grid is on the peak shaving threshold is shown in figure 8.9. This is the most appropriate use of the battery, as peak shaving generates by far the greatest financial profit. To find out if a battery on 300 kWh is more climate friendly than the current battery, the total reduction in GHG emissions from 2015 is used. The result is 750 kg

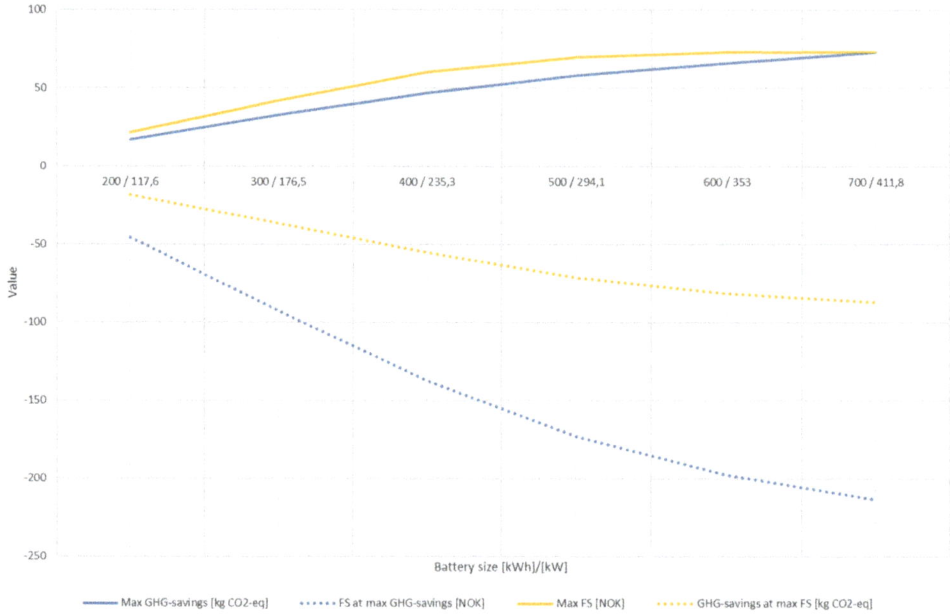

Figure 8.9. Comparison of various battery sizes for use on campus Evenstad when maximum import is on the peak shaving threshold. Data from the period 27 February–5 March 2015.

of CO_2. Over the battery's lifetime it will compensate for 11 of a total 52 tons of CO_2-equivalents in built-in emissions, making the net emissions increase by 11 tons compared to the current battery of 204 kWh. Because the climate savings decrease gradually from 300 kWh to 700 kWh and the built-in emissions to the battery are linear, it is not considered climate friendly to increase the size of the battery on campus Evenstad.

If the current battery did not have the limitations made by the UPS, allowing the whole capacity to be utilized, reductions in GHG emissions would have been 610 kg CO_2-equivalents in 2015. Over the battery's lifetime it would have compensated for 9.2 tons, which equalise 26% of the battery's climate debt.

8.4 Discussion

The results show that the Norwegian electricity market is not ideal for use of residential batteries. By connecting the Norwegian power market closer to the continent, the spot price can increase or alternate more during the days, and there is a possibility for larger profit. However, the price and carbon intensity will still have an inverse correlation. Therefore, the gain will be questionable. Furthermore, neither new batteries, new batteries produced on renewable energy, nor refurbished batteries can compensate for embodied emissions in this application under the current Norwegian electricity conditions and chosen assumptions. However, refurbished batteries should definitely be explored further, as the total climate impact may prove substantially lower than producing a new battery for the purpose, and repurposing a battery is significantly cheaper than recycling it.

The limitations of this study are still rather significant, as the computational power and programming required to have a thorough model were not available. For this reason, the data set for carbon intensities and spot price modelling were focused only on one week from 2015. This causes a bias when extrapolating this data to a year period.

8.4.1 Maximizing self-consumption of PV energy

The economic gain by maximizing self-consumption of locally produced energy is small, as stated in the results. Especially in a local perspective, where the grid energy generally has a lower carbon intensity than the energy produced from the PVs. In a global perspective, the prospects are somewhat different. The European average electricity mix has a carbon intensity of approximately 300 g per kWh, which is considerably higher than the 50 g per kWh that represent the PVs at campus Evenstad. Even though the emissions will increase locally, this will slightly reduce the emissions on a European scale.

8.4.2 Peak shaving

All of the models assume constant values for import and export for an hour. This is somewhat imprecise, as the power can alternate greatly within the hour, which can affect the battery. In reality, the campus could be in need of importing power in times where the model tells otherwise. This can potentially lead to the battery bank not being able to meet the demand.

The potential for peak shaving varies between 36.80 and 58.26 kWh per h per year and creates a difference of almost 12 000 NOK per year. Therefore, it is hard to set a recommended threshold for peak shaving. If the battery bank connects at low values, it may be empty when it is really needed. If it connects too late, its full potential will not be exploited.

Peak shaving is vulnerable for long periods of high-power use. For example, a day with construction work on campus can lead to exceptionally high-power usage, risking the battery bank emptying in the middle of the power peak. The battery bank will then be useless. A solution to periods with higher expected power usage can be to rent a bio-diesel aggregate to help generate power.

The climate consequences of peak shaving are small. Peak shaving happens only a few hours per year and is usually during daytime. The biggest climate factor will be the losses due to efficiency. As the battery is used so seldom for peak shaving, it is compatible with saving emissions by exploiting the alternations in the carbon intensity in the grid.

8.4.3 The value of a battery at Campus Evenstad

The battery's main purpose is UPS, and that prevents the full exploitation of the battery's potential. However, it turns out that even without UPS, the battery does not give positive results. As every use of battery will increase both GHG emissions and costs, there is no value in upscaling the battery bank. As the current battery is neither environmentally nor economically beneficial, the battery could have been limited to the UPS capacity of 100 kWh.

It appears that the alternative that generates least losses is to sell the excess half of the battery. Then the campus would escape substantial economical burdens. If there is a desire to keep the current battery bank for research purposes, the most appropriate use is a combination of peak shaving and exploiting fluctuations in carbon intensity to reduce emissions. The two models work well together, as the battery is fully charged during daytime, awaiting peak shaving and is also used to supply the campus with clean energy during the night-time. Even though this is the cleanest alternative, the abatement cost is still relatively high.

References

[1] Ellingsen L A, Majeau-Bettez G, Singh B, Srivastava A K, Valøen L O and Strømman A H 2014 Life cycle assessment of a lithium-ion battery vehicle pack *J. Ind. Ecol.* **18** 113–24

[2] Eddahech A, Briat O and Vinassa J-M 2015 Performance comparison of four lithium-ion battery technologies under calendar aging *Energy* **84** 542–50

[3] Eckart J 2017 Batteries can be part of the fight against climate change—if we do these five things *World Economic Forum*

[4] Kim H C, Wallington T J, Arsenault R, Bae C, Ahn S and Lee J 2016 Cradle-to-gate emissions from a commercial electric vehicle Li-ion battery: a comparative analysis *Environ. Sci. Technol.* **50** 7715–22

[5] Bobba S *et al* 2018 Life cycle assessment of repurposed electric vehicle batteries: an adapted method based on modelling energy flows *J. Energy Storage* **19** 213–25

[6] Faria R *et al* 2014 Primary and secondary use of electric mobility batteries from a life cycle perspective *J. Power Sources* **262** 169–77

[7] Sun S I, Crossland A F, Chipperfield A J and Wills R G A 2019 An emissions arbitrage algorithm to improve the environmental performance of domestic PV-battery systems *Energies* **12** 560

[8] NVE Nasjonal varedeklarasjon 2019 [Internet] 2018 [cited 2019 Jun 26]. Available from: https://nve.no/energiforsyning/varedeklarasjon/nasjonal-varedeklarasjon-2018/

[9] Hovland K M 2018 Strømavgift kan gi batteriboom

[10] Clauß J, Stinner S, Solli C, Lindberg K B, Madsen H and Georges L 2018 A generic methodology to evaluate hourly average $CO_{2eq.}$ intensities of the electricity mix to deploy the energy flexibility potential of Norwegian buildings *Proc. of the 10th Int. Conf. on System Simulation in Buildings, Liege, Belgium* pp 10–2

[11] Jensen S, Opsahl E and Monsen M 2018 Batterier for norske boliger og næringsbygg *NTNU*

[12] Klimakur 2010 Tiltak og virkemidler for å nå norske klimamål mot 2020. TA2590/2010, Regjeringen

[13] Gjertsen M 2019 Optimal control of batteries and hot water heaters for energy flexibility in zero emission neighbourhoods *NTNU*

[14] NordPool 2015 Elspot Prices 2015 Hourly NOK

www.ingramcontent.com/pod-product-compliance
Ingram Content Group UK Ltd.
Pitfield, Milton Keynes, MK11 3LW, UK
UKHW051850100926
8556IPUK00005B/553

9 780750 332576